International Conference for Free and Open Source Software for Geospatial (FOSS4G 2016)

OSGeo Journal Volume 16, Issue 1

Bonn, Germany
24-26 August 2016

Editors:

Franz-Josef Behr

ISBN: 978-1-7138-1902-8

Printed from e-media with permission by:

Curran Associates, Inc.
57 Morehouse Lane
Red Hook, NY 12571

Some format issues inherent in the e-media version may also appear in this print version.

OSGeo
9450 SW Gemini Dr. #42523
Beaverton, Oregon
United States
97008

www.osgeo.org

Additional copies of this publication are available from:

Curran Associates, Inc.
57 Morehouse Lane
Red Hook, NY 12571 USA
Phone: 845-758-0400
Fax: 845-758-2633
Email: curran@proceedings.com
Web: www.proceedings.com

TABLE OF CONTENTS

IDENTIFICATION OF SAR DETECTED TARGETS ON SEA IN NEAR REAL TIME APPLICATIONS FOR MARITIME SURVEILLANCE 1
Sergey Voinov, Egbert Schwarz, Detmar Krause, Matthias Berg

BUILDING APPLICATIONS WITH FOSS4G BRICKS: TWO EXAMPLES OF THE USE OF GRASS GIS MODULES AS A HIGH-LEVEL "LANGUAGE" FOR THE ANALYSES OF CONTINUOUS SPACE DATA IN ECONOMIC GEOGRAPHY 11
Moritz Lennert

INVOLVING COMMUNITIES IN ENVIRONMENTAL PROTECTION BY COMMUNITY INFORMATION SYSTEMS: THE CASE STUDY OF "LA CUICADORA" 23
Domenico Vito

IS THERE LIFE IN VIRTUAL GLOBES? 31
Konstantinos Evangelidis, Theofilos Papadopoulos

FROM GLOBAL OBSERVATIONS TO LOCAL INFORMATION: THE EARTH OBSERVATION MONITOR 39
Jonas Eberle

MODELLING POPULATION INTERACTIONS: THE PYSAL AND OPEN SOURCE WAY 56
Lorraine Barry

SHAPEFILE PROJECTION FINDER 59
Manfred Egger

OPEN SOURCE GEOSPATIAL APPLICATIONS TO SUPPORT RIVER BASIN MANAGEMENT IN KHARAA RIVER BASIN, MONGOLIA 62
Jürgen Hofmann, Vanessa Bremerich, Jonas Gröning, Marc Janssen, Hinrich Paulsen

CLASSIFICATION OF LAND USE/COVER CHANGE IN THE KANTO REGION, JAPAN USING FOSS4G AND OPEN DATA 65
Nobusuke Iwasaki, David Spraguea, Ayaka Onohara

MODELLING CONCRETE AND ABSTRACT GEOPROCESSING WORKFLOWS WITH ILWIS 4 68
Rob Lemmens

HOW TO CROSS A SQUARE? 71
Jakob Miksch, Stefan Hahmann

THE CONSTRUCTION OF OPEN DATA PORTAL USING DKAN FOR INTEGRATE TO MULTIPLE JAPANESE LOCAL GOVERNMENT OPEN DATA 74
Toshikazu Seto, Yoshihide Sekimoto

DEVELOPMENT OF ENVIRONMENT SPATIAL INFORMATION SERVICE UTILIZING MOBILE LOCATION-BASED TECHNOLOGY 76
Jeongho Yoon

GRASS IN THE DESERT? DEVELOPING FOSS TOOLS FOR MONITORING DESERTIFICATION 79
Anna Zanchetta

Author Index

Free and Open Source Software for Geospatial (FOSS4G) Conference Proceedings

Volume 16 *Bonn, Germany* Article 20

2016

Volume Introduction Letter

Franz-Josef Behr
Stuttgart University of Applied Sciences

Volume Introduction Letter

This Conference Proceedings is a collection of selected papers and posters submitted to the Academic Program of the International Conference for Free and Open Source Software for Geospatial (FOSS4G 2016), 24th to 26th August 2016 in Bonn, Germany. Again geospatial open source software was the focus of a well-attended international conference flanked by presentations on open standards and the open data movement, a meeting place for open communities.

Similar to previous FOSS4G conferences at the national and international level, the academic papers and posters cover an extensive wide range of topics reflecting the contribution of the academia to this field by the development of open source software components, in the design of open standards, in the proliferation of web-based solutions, in the dissemination of the open principles important in science and education, and in the collection and the hosting of freely available geo-data.

This volume follows a 10-year tradition of FOSS4G Academic Proceedings that with many published on the former "OSGeo Journal" website. However, with the help of the University of Massachusetts Amherst Scholarly Communication team, we are publishing this volume and individual papers using the new bepress-supported publishing system. A benefit of this system is that it feeds many library databases making the proceedings and individual papers much more identifiable through library search mechanisms.

On the page *https://scholarworks.umass.edu/foss4g/vol16/iss1/* the reader will get an overview about the selected papers and posters published in the Conference Proceedings.

A few additional papers found their publication on other platforms. The paper "PaleoMaps: SDI for open paleoenvironmental GIS data"[1] by Christian Willmes, Daniel Becker, Jan Verheul, Yasa Yener, Mirijam Zickel, Andreas Bolten, Olaf Bubenzer, Georg Bareth and the paper "Development of a New Framework for Distributed Processing of Geospatial Big Data"[2] by Angéla Olasz, Binh Nguyen Thai, Dániel Kristóf have been published through the *International Journal of Spatial Data Infrastructures Research* of the European Commission's Joint Research Centre, organized by Sven Schade. Anita Graser's paper "Towards landmark-based instructions for pedestrian navigation systems using OpenStreetMap"[3] has been published in the *AGILE 2017* proceedings.

The presentations of the conference are also available in the AV-Portal of TIB, the German National Library of Science and Technology[4]. All videos are assigned a digital object identifier (DOI) and a media fragment identifier to cite presentations as a whole or single segments separately.

The support of Christian Willmes and Sergey Voinov amongst other reviewers is gratefully acknowledged. Moritz Lennert pushed the process forward. The support of Charles Schweik, Mohammed Zia and Erin Jerome was very helpful and encouraging!

As the editor of the FOSS4G 2016 proceedings I'd like to encourage the Academia to intensify the usage of open software, tools, data, science and education and to contribute actively to forthcoming conferences on local, national and international level searching the contact and exchange with practitioners, developers and users to extend and foster the worldwide community – "Building bridges", as the motto of AGSE 2016 was.

Sincerely,

Franz-Josef Behr – FOSS4G 2016 Academic Committee chair and Proceedings editor

---

[1] http://ijsdir.jrc.ec.europa.eu/index.php/ijsdir/article/view/431

[2] http://ijsdir.jrc.ec.europa.eu/index.php/ijsdir/article/view/423

[3] https://agile-online.org/images/conference_2017/Proceedings2017/shortpapers/46_ShortPaper_in_PDF.pdf

[v] Via https://av.tib.eu/series/253/foss4g+bonn+2016

Free and Open Source Software for Geospatial (FOSS4G) Conference Proceedings

Volume 16 *Bonn, Germany* Article 1

2016

Identification of SAR Detected Targets on Sea in Near Real Time Applications for Maritime Surveillance

Sergey Voinov
German Aerospace Center (DLR) – German Remote Sensing Data Center (DFD

Egbert Schwarz
German Aerospace Center (DLR) – German Remote Sensing Data Center (DFD)

Detmar Krause
German Aerospace Center (DLR) – German Remote Sensing Data Center (DFD)

Matthias Berg
German Aerospace Center (DLR) – German Remote Sensing Data Center (DFD)

Recommended Citation

Voinov, Sergey; Schwarz, Egbert; Krause, Detmar; and Berg, Matthias (2016) "Identification of SAR Detected Targets on Sea in Near Real Time Applications for Maritime Surveillance," *Free and Open Source Software for Geospatial (FOSS4G) Conference Proceedings*: Vol. 16 , Article 1.
DOI: https://doi.org/10.7275/R5125QTW
Available at: https://scholarworks.umass.edu/foss4g/vol16/iss1/1

Identification of SAR Detected Targets on Sea in Near Real Time Applications for Maritime Surveillance

Sergey Voinov*, Egbert Schwarz, Detmar Krause, Matthias Berg

German Aerospace Center (DLR) – German Remote Sensing Data Center (DFD) –
sergey.voinov|egbert.schwarz|detmar.krause|matthias.berg@dlr.de

KEYWORDS: Remote sensing, Synthetic Aperture Radar (SAR), Automatic Identification System (AIS), data fusion, Maritime Surveillance, Maritime Situational Awareness.

ABSTRACT

Remote sensing technologies are widely used in maritime surveillance applications. Nowadays, spaceborne Synthetic Aperture Radar (SAR) systems provide outstanding capabilities for target detection at sea for large areas independently from the weather conditions. The generated value added target detection product is composed by complementary information from the Automatic Identification System (AIS). Resulting information layers provides a more reliable picture on the maritime situation awareness. This paper describes the approach of SAR-AIS data fusion and its visualization means developed for Near Real Time (NRT) Applications for Maritime Situational Awareness by the Maritime Security Lab at the Ground Station in Neustrelitz, part DLR's German Remote Sensing Data Center (DFD). Presented implementation is based on combination of many open source geospatial libraries and frameworks (e.g., GDAL/OGR, Geoserver, PostgresSQL) and shows their effectiveness in the context of complex automated data processing in the frame of NRT requirements.

* Corresponding author

1 Introduction

The global economy is highly dependent on the shipping industry. Because of maritime transport development, nowadays ocean does not divide the world, but connects. Shipping has always been the only low-cost method of freight transport over long distances. Today, 80% of all goods are transported over the sea including oil, oil products, coal, ore, grains and others. With growing world's economy the exchange of goods utilizing maritime transport on the global scale is growing steadily, which causes an increase of ship traffic. New routes are being developed, e.g., –a North -East or North- Wert passage, opening new routing opportunities due to the global climate changes. To understand the situation and ensure maritime safety and security regarding to the shipping density as well as environmental monitoring, maritime surveillance applications are nowadays of paramount importance.

Spaceborne Synthetic Aperture Radar (SAR) systems have already proven their effectiveness on maritime surveillance. The main advantage of SAR sensors is their capability to perform observations of the Earth's surface independent from the weather conditions and time of the day. Furthermore, they can cover large areas – up to several hundreds of kilometers wide at the same time. Limitations related to the revisit time are still there, but could be reduced by involving several different missions, e.g., – TerraSAR-X, Radarsat-2, and Sentinel-1. Data acquired from these satellites demonstrated its efficiency in maritime surveillance applications, including such task as target detection at sea.

Although the SAR target (ship) detection product provides ship's position and its parameters, such as width, length and even heading (Tings, et al., 2015), this information can still not provide a complete maritime picture. Complementary information from vessel tracking systems, in particular from Automatic Identification System (AIS) can provide such parameters, like vessel type (according to international classification), speed, course over ground as well as its width and length. AIS messages are broadcasted with an update rate up to some seconds, and mandatory for all vessels larger than 300 of gross tons and all passenger ships irrespective of size.

Integration of both datasets, – SAR-derived detections and AIS data, enables to perform much more complex analysis for maritime surveillance. In particular, SAR detection would confirm vessel's physical persistence in addition to AIS reports. At the same time, vessels can be detected which due to technical reasons or malicious actions not broadcasting AIS messages.

The overall state of the NRT applications in the maritime domain at DLR's Ground Station Neustrelitz, is given in (Schwarz, et al., 2015). This paper emphases on the complete automated processing chain of deriving ship detection products with associated AIS information, which already has been integrated for operational use and being under further development.

The paper is organized as follows. Section 2 gives an overview about the whole processing chain. Section 3 explains the approach of SAR-AIS data fusion. The dissemination and visualization of resulting products are described in Section 4.

2 Processing, Workflow and Framework

Ground Station Neustrelitz is the main receiving and processing facility for the TerraSAR-X mission composed by the TerraSAR-X and TanDEM-X satellites. In 2015 the capabilities of the Ground Station has been extended with a Regional Ground System (RGS) to support Radarsat-2. Furthermore the collaborative ground segment to support the Copernicus Mission Sentinel-1 has been developed.

The data, received by the ground station are immediately transferred to the processing system. The whole processing is controlled by the Processing System Management (PSM) developed by DLR and Werum Software & Systems AG (Boettcher, et al., 2001). The implemented PSM control system forces rule based scheduling of different processors and call them step-by-step or in a parallel sequence. Figure 1 shows the general workflow sequence of the ship detection NRT application controlled by the PSM.

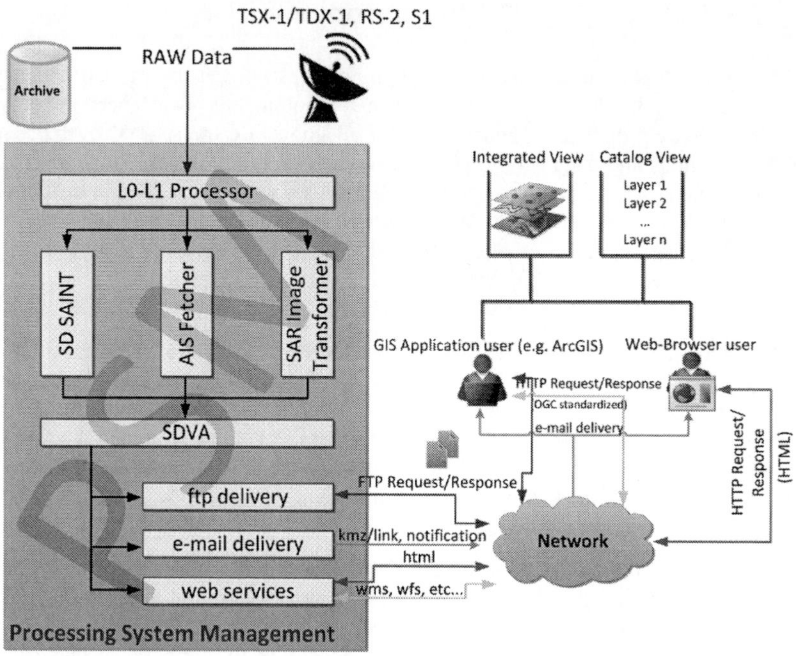

Figure 1: NRT ship detection processing workflow

The first value adding processor requires satellite sensor specific processing up to level 1b (L1b). At this point, RAW sensor data being processed to the image form with applied radiometric calibrations and extracting geolocation information. For TerraSAR-X and TanDEM-X satellites - the TerraSAR Multi Mode SAR Processor (TMSP) takes care on this task (Breit, et al., 2008).

The next three processing steps, SD SAINT; AIS Fetcher and SAR Image Transformer are scheduled in parallel as they are independent from each other. All three processor modules using the same L1b input product, and they are common for all available SAR sensors.

For the SAR target detection task the SAR AIS Integrated Toolbox (SAINT), developed in the Maritime Safety and Security Lab in Bremen, is used. The detection algorithm based on constant false alarm rate (CFAR) detector. Then, the derived binary mask, showing positions of ships on the image, is used for the parameter estimation of every detected vessel. More detailed information about the algorithms and methods used in SAINT are described in (Tings, et al., 2015) and (Brusch, et al., 2010). The output of the SAINT is a point data layer showing vessel positions in projected geographical coordinates with estimated attributive values like vessel's length, width and heading.

AIS Fetcher performs collection of the AIS data for required time extent (imagery time) and coordinates (geographical bounding box of the input L1b image). The processor has been developed in the Maritime Security Lab Neustrelitz, part of the DFD. It queries the AIS data from local database (DB) in case of historical requests or collects them from external sources like AIS providers. The software is also capable to decode RAW AIS sensor data in several formats and push them into the local DB for future use. The output of AIS Fetcher is a point data set with dynamic and static AIS attributes. Every sequence of AIS messages for specified request is processed through the embedded AIS Plausibility Processor (Heymann, et al., 2013), developed in the Institute of Communications and Navigation, part of DLR. The Processor checks certain values of the AIS message data in the context of the plausibility of the vessel behavior, in particular how fast the vessel moves, how it changes the course, etc. The processor extends every AIS message with additional fields showing possible errors in certain attributes.

The image processing software SAR Image Transformer is called to apply automatically geolocation information provided with L1b image, adjust the histogram and generates several outputs: The full resolution L1b geotiff image in UTM and in EPSG:4326 projections. Currently the processor supports data processing from the satellites TerraSAR-X, Radarsat-2 and Sentinel-1. Furthermore, quicklook products in .png and .kmz format (for visualization in Google Earth) are generated.

The next, according to this this paper most relevant, processing step is the data fusion of SAR detected ships and AIS reports. The Ship Detection Value Adder (SDVA) developed in the Maritime Security Lab Neustrelitz executes the data fusion operation. As an input the SDVA accepts the output from all the previous processing steps. Namely, SAR ship detection results (point data), AIS reports for specified area and time (point data), the quicklook image (for visualization of the results), and the L1b image metadata. The algorithm behind will be discussed in the next section. The resulting data fusion product is generated in different final formats like Google kml/ kmz, ESRI shape file, json and many others.

The last step of the NRT ship detection chain handles the product dissemination. Different delivery options are available like ftp/ sftp, via e-mail (e.g. - kmz), as well as via web mapping services, enabling customers the direct use of product results in their GIS applications over OGC interfaces (wms, wfs) or in web-browser using web-mapping client.

3 SAR-AIS Data Fusion

The Ship Detection Value Adder (SDVA) performs the data fusion between SAR detected targets and AIS reported information for a requested spatial and time extent. The general workflow of SDVA is shown the figure 2.

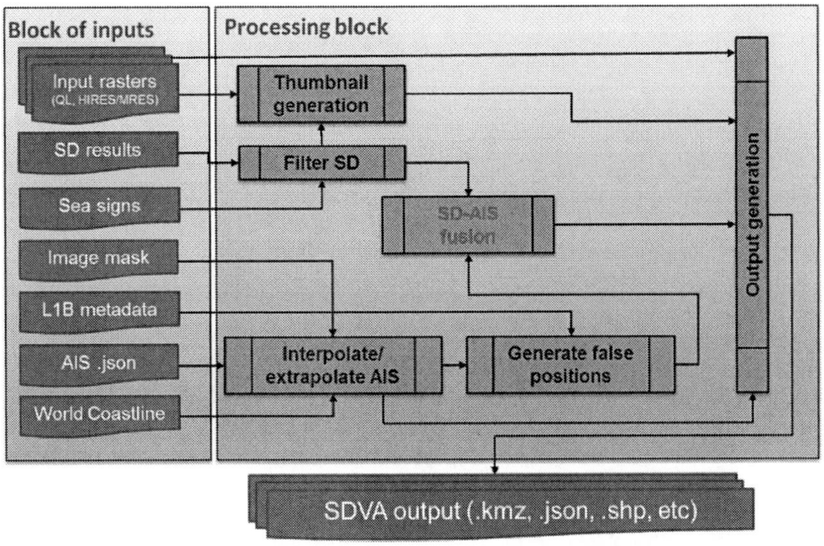

Figure 2: Ship Detection Value Adder workflow.

In order to fulfill near real time delivery requirements, most of the operations in SDVA, by analogy to all the other processors of NRT ship detection chain, are implemented in a high level of parallelization.

Nowadays, besides the ships, there are many generally known objects on the open sea. In particular, these are windfarms, oil platforms and buoys. Such objects are often detected as vessels, and looking on their estimated parameters, they are really not much different from the real detected vessels. Therefore a filtering step of ship detection results has been implemented. Filtering is done by checking if the detected object spatially intersects with objects in the database of sea signs. Once filtering is done, a thumbnail (zoomed in picture from original geotiff) will be generated for the final result. These thumbnails are embedded into the kmz product file and WMS services for the OGC feature info requests.

3.1 AIS Interpolation

The L1b image will be available within 10 minutes after downlink. Because of this delay, it is possible to collect some AIS data not only for the time before the image acquisition, but also up to 10 minutes afterwards. This allows reconstructing of vessel tracks in order to get an AIS ship position of highest accuracy according to the imagery time. The implemented solution enables more precise SAR-AIS data fusion based on the exact imagery time for every ship and solve the following problems:

1) In the real world the AIS reports for moving vessels are being broadcasted in a different reporting interval. The transmission of updates mainly depends on the speed over ground (SOG) of the vessel. Furthermore updates occur at different intervals depending on whether a Class A or Class B transponder is used. Unfortunately, not always all of transmitted reports are being collected due to the different technical reasons. Furthermore, often

AIS providers resample the data by time. As the result, in some cases the update rate for available AIS reports could be up to a few minutes.

2) By using different acquisition modes SAR images are extended to cover very large areas, up to thousands of square kilometers. To scan such big area SAR sensor needs several seconds up to more than one minute. Even 30 seconds make big sense, e.g., a cargo vessel moves on the open sea with the speed of 20 knots which is around 10 meters per second. During the only 30 seconds, this ship will pass the distance of about 300 meters. Therefore, for prediction of the AIS position at imagery time, a specific imagery time in the proper image area should be considered. The auxiliary information for time calculation in specific places on the image is available in L1b metadata.

The first step is a reconstruction of AIS tracks with minimum specified time step. The new calculated positions are derived using "dead reckoning" concept, based on the last position, speed and course over ground as illustrated on figure 3.a.

The new positions are always crosschecked if they are not intersecting the coast line. If the new position falls on the land area, it is iteratively corrected as shown on Figure 3.b.

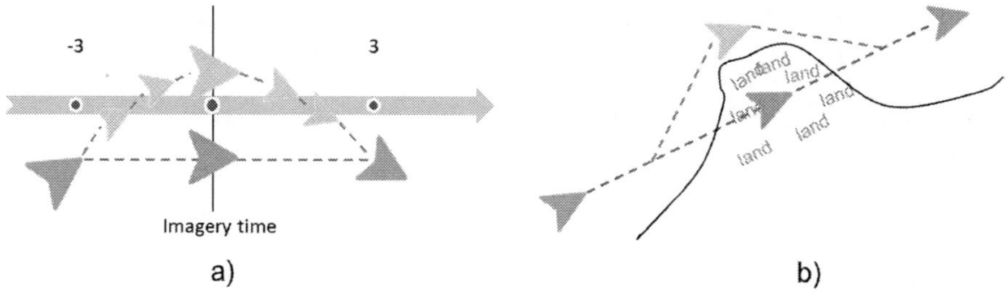

7

Figure 3: AIS track reconstruction

The new attribute values for speed and course over ground are calculated by local weighted interpolation from known surrounding points

$$a_i = a_{i-1}w_{i-1} + a_{i+1}w_{i+1} \qquad (1)$$

where a_i is the attribute value for a new point to be calculated, a_{i-1} and a_{i+1} are the attribute values from neighboring points, w_{i-1} and w_{i+1} are corresponding weights for the known attributes and based on their distance from predicted point:

$$w_{i-1} = 1 - \frac{d_{i-1}}{d_{i-1}+d_{i+1}} \quad \text{and} \quad w_{i+1} = 1 - \frac{d_{i+1}}{d_{i-1}+d_{i+1}} \qquad (2)$$

where d_{i-1} is the distance from "left" known point to predict, and $d_{i+1}$is the distance from predicted point to the "right" known point.

If the predicted point is outside of the known track, namely has only one known neighbor, then the last known attribute is taken as the new point.

AIS positions at imagery time are derived in the following way. The reference time for the whole image must be defined. The optimal choice is to take a middle time point between image start and stop times. Then, it is necessary to find the closest AIS reports in time, relative to the reference time. In the next step, AIS geographical coordinates that reported closest to the reference time are used to determine on which part of the image they would appear. Normally, for every L1b SAR image the metadata provides means for computation the imaging time for every pixel. Furthermore, the time deviation in range (dimension x) direction is not significant; the most important is the time according to the azimuth (dimension y) direction. This can be calculated in different ways: out of provided georeferenced grid or by simple linear interpolation between imaging start and stop times.

3.2 Project Displaced Positions

Dealing with SAR images it is known that moving targets can be displaced on the image from their actual position. This effect caused by additional Doppler shift introduced by moving objects. If the object is moving in the flight direction (along-track) with the satellite, then it will be just blurred. When the object motion trajectory is crossing the sensor (across-track), the object's position will be shifted in the azimuth direction. Depending on the object's velocity, such displacement can reach up to several hundred meters. Therefore, prior the fusion of SAR detected targets and AIS reports, this effect has to be considered. More detailed explanation of theory of moving objects in SAR images can be found in (Meyer, et al., 2007). Some examples of azimuth displacement are shown on Figure 4.

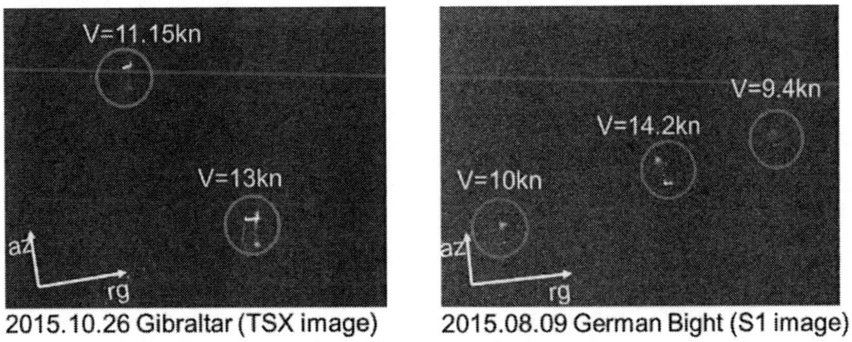

Figure 4: Azimuth displacement due to across-track movement

The azimuth shift of the moving object can be estimated following way

$$\Delta_{az} = -R\frac{V_{at}}{V_{sat}}[m]$$

where Δ_{az} is object's azimuth displacement, R is the distance between the object and the satellite, V_{sat} is the satellite velocity and V_{at} is the object velocity in across-track direction.

For several decades, researchers have studied object motions in SAR images. A number of methods to compensate this effect have been proposed, (e.g., Raney, 1971; Meyer, et al., 2005; Arii, 2014). In particular, with integration of a priori knowledge, such like road databases and vehicles radar signatures (Meyer, et al., 2007), or by direct velocity estimation out of image characteristics (Raney, 1971; Arii, 2014).

However, AIS reports provide the information about vessel's speed and course, which offer the possibility to calculate V_{at} and Δ_{az} respectively. The reversed method was applied, where for AIS reports at imaging time a displaced position on the image is calculated. This approach is of low computational cost which is crucial for NRT services. Although the errors are possible, especially on port areas, it provides sufficient accuracy and widely used in the context of SAR-AIS data fusion, e.g., (Zhi Zhao et al., 2014; Mazzarella, et al., 2015).

3.3 SAR-AIS Data Fusion

After extracting AIS reports at imaging time and projecting their positions according to the Doppler shift a fusion between AIS reports and SAR-detections takes place.

SAR-AIS composition is implemented by double nearest neighbor search. First, for every AIS message with projected position on the image a nearest SAR detected target has to be associated. For different reasons, e.g., low resolution of SAR image it is possible that for a particular AIS report, there is no corresponding SAR target detection. Therefore a special, empirical threshold for the search area was specified. Once the closest SAR target has been found, the distance (Das) between AIS and SAR objects is remembered. If there was nothing found within the searching area, it is assumed there is no target for assigning.

The second step is iteration through the SAR objects and linking to the closest AIS report. This operation is similar to the AIS-SAR relation described above. The distances between SAR and AIS targets (Dsa) are saved as well.

Comparison of AIS-SAR and SAR-AIS relations is applied before finishing this process. When both relations in both directions have the same distances (Das = Dsa), it is assumed the association is valid, and attributes from AIS are linked to the corresponding SAR target.

4 Visualization

Providing NRT services it is very important that the end user gets the resulting products in time. Besides skilled GIS analysts, the data should be readable and comprehensible for the customer and require no special software. This concept was the determining factor in designing the products. For this purpose a number of output formats and delivery methods are implemented. As shown in Figure 1, currently three possible delivery methods are implemented:

1 Transfer files via sFTP or FTP server. This method is oriented for experienced GIS analysts, which prefer to define their own visualization and perform deeper analysis. After the processing, files are automatically transferred to either DLR delivery server or user site. The file set may include: ESRI shape files (as well as json and ASCII) containing ship detection results (point data) with assigned AIS attributes; layer with all available AIS data for the scene; quicklook or full resolution SAR image in geotiff or png format.

2 E-mail notification – a short message notifying about available results via FTP (described above) and set of files attached, with the maximum total size of 5 MByte. The e-mail could contain a set of different files based on the user requirements, e.g., an ASCII text file with detected targets, a georeferenced quicklook png image in UTM projection, as well as a kmz file. The kmz file is composed by integrated layer view of results, including the SAR ship detections, AIS reports (points and restored tracks) and quicklook overlay of SAR image. Example of kmz is shown on figure 5.

3 The web services solution enables dissemination of NRT products to the user without copying them (in the usual meaning of copy). Two ways of using web mapping services are possible: direct integration into the GIS software over the OGC services (wms, wfs), served by GeoServer, or using web-mapping client, which requires only a web browser. The web-mapping client is based on front-end components of Environmental and Crisis Information System (UKIS), developed at DFD, which is implemented by using the latest web technologies, such as AngularJS and Leaflet. The client is platform independent and supports any modern web-browser without need of any extra software or plugins. Example of product view in the UKIS client shown in Figure 6.

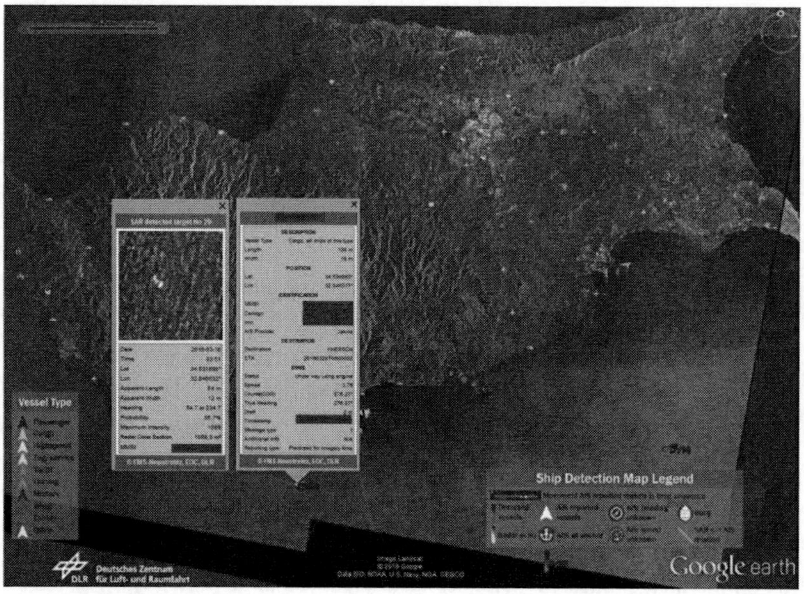

Figure 5: Example of kmz output visualized in Google Earth

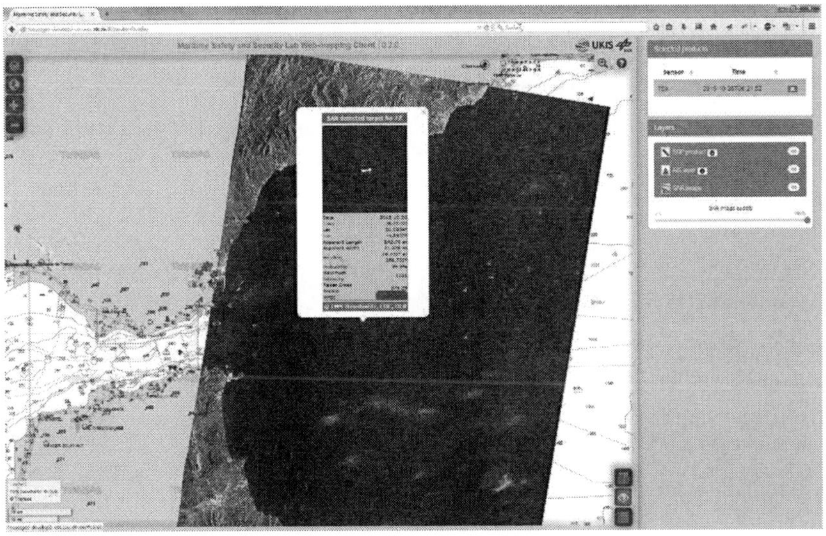

Figure 6: Product view in web-mapping client

5 Conclusion

This paper discusses the SAR-AIS data fusion methods, part of the automated processing chain for NRT ship detection and identification, developed at DLR. The workflow is implemented in a high level of parallelization, to ensure deriving ship detection products within 10 to15 minutes after image acquisition.

Although the presented approach showed good results according to the service performance and product accuracy, regardless there is still space for improvements. In particular, for AIS interpolation and data fusion more complex models could be applied in order to improve the accuracy, especially near the coast or port areas where the vessel routes are not linear. This can be improved, for example, by utilizing additional auxiliary information about the traffic on the sea like it was proposed by (Mazzarella, et al., 2015).

The integration of the new sensors and data streams, not only from radar, but also optical sensors, as well as satellite based AIS will increase the reliability of the service.

6 References

Arii, M. (2014), 'Efficient Motion Compensation of a Moving Object on SAR Imagery Based on Velocity Correlation Function.' in IEEE Transactions on Geoscience and Remote Sensing, 52(2), 936-946.

Boettcher, M., Reißig, R., Mikusch, E., & Reck, C. (2001), 'Processing Management Tools for Earth Observation Products at DLR-DFD.' in 'Procedings of DASIA 2001 Data Systems in Aerospace. (SP-483. Nice, 2001)', European Space Agency.

Breit, H., Schättler, B., Fritz, T., Balss, U., Damerow, H., & Schwarz, E. (2008), 'TerraSAR-X Payload Data Processing: Results from Commissioning and Early Operational Phase.', in IEEE Geoscience and Remote Sensing Symposium (IGARSS), (pp. 209-212), Boston, USA.

Brusch, S., Lehner, S., Fritz, T., M., S., Soloviev, A., & van Schie, B. (2011), 'Ship Surveillance With TerraSAR-X', in IEEE Transactions on Geoscience and Remote Sensing, 49, 1092-1103. doi:10.1109/TGRS.2011.2117471

Heymann, F., Noack, T., & Banyś, P. (2013), 'Plausibility analysis of navigation related AIS parameter based on time series.' in 'Proceedings of European Navigation Conference. (Vienna, 2013).

Mazzarella, F., Vespe, M., & Santamaria, C. (2015), 'SAR Ship Detection and Self-Reporting Data Fusion Based on Traffic Knowledge.', in IEEE Geoscience and Remote Sensing Letters, 12(8), 1685-1689.

Meyer, F., Hinz, S., Laika, A., & Bamler, R. (2005), 'A-priori Information Driven Detection of Moving Objects for Traffic Monitoring by SAR.', in IEEE International Geoscience and Remote Sensing Symposium, 4, pp. 2932-2936.

Meyer, F., Hinz, S., Müller, R., Palubinskas, G., Laux, C., & Runge, H. (2007), 'Towards traffic monitoring with TerraSAR-X.', in Canadian Journal of Remote Sensing, 33, 39-51.

Raney, R. (1971, May). 'Synthetic aperture imaging radar and moving targets.' in IEEE Transactions on Aerospace and Electronic Systems, AES-7(3), 499-505.

Schwarz, E., Krause, D., Daedelow, H., & Voinov, S. (2015), 'Near Real Time Applications For Maritime Situational Awareness.', 'Proceedings of Deutscher Luft- und Raumfahrtkongress. (Rostock, 2015).'

Tings, B., Bentes da Silva, C. A., & Lehner, S. (2015), 'Dynamically adapted ship parameter estimation on TerraSAR-X images.' in International Journal of Remote Sensing. doi:10.1080/01431161.2015.1071898

Zhi Zhao, K. J. (2014), 'Ship Surveillance by Integration of Space-borne SAR and AIS – Further Research.' in The Journal of Navigation, 67, 295-309. doi:10.1017/S0373463313000702

Free and Open Source Software for Geospatial (FOSS4G) Conference Proceedings

Volume 16 *Bonn, Germany* Article 3

2016

Building applications with FOSS4G bricks: two examples of the use of GRASS GIS modules as a high-level "language" for the analyses of continuous space data in economic geography

Moritz Lennert

Department of Geosciences, Environment and Society, Free University of Brussels (ULB)

Building applications with FOSS4G bricks: two examples of the use of GRASS GIS modules as a high-level "language" for the analyses of continuous space data in economic geography

Moritz Lennert [a, *]

[a] Department of Geosciences, Environment and Society, Free University of Brussels (ULB) – moritz.lennert@ulb.ac.be

KEY WORDS: GRASS GIS, economic geography, spatial modeling, raster analysis, programming

ABSTRACT:
In a world where researchers are more and more confronted to large sets of micro- data, new algorithms are constantly developed that have to be translated into usable programs. Modular GIS toolkits such as GRASS GIS offer a middle way between low-level programming approaches and GUI-based desktop GIS. The modules can be seen as elements of a programming language which makes the implementation of algorithms for spatial analysis very easy for researchers. Using two examples of algorithms in economic geography, for estimating regional exports and for determining raster-object neighbourhood matrices, this paper shows how just a few module calls can replace more complicated low-level programs, as long as the re- searcher can change perspective from a pixel-by-pixel view to a map view of the problem at hand. Combining GRASS GIS with Python as general glue between modules also offers options for easy multi-processing, as well as supporting the increasingly loud call for open research, including open source computing tools in research.

[*] Corresponding author

1 Introduction

Science is constantly evolving and new indicators and analyses have to be developed. All to often, researchers reinvent the wheel by programming their algorithms in low-level code. This can possibly make this code very machine efficient if it is coded correctly, but it can also lead to a waste of time and energy that could better be spent on the content instead of the programming. Obviously many efforts to wrap existing, and well-tested, routines into libraries and other forms of packages exist, and most researchers use those. In spatial sciences, however, it seems that re- searchers often either use very high-level GUI-oriented desktop analysis systems, or they use fairly low-level systems, such as, for example, the fiona and rasterio Python libraries or their equivalent in other programming languages.

In the FOSS4G ecosystem, there is a middle species, though: tools that are composed of multiple command-line modules that can be chained to form very efficient work processes. In this approach each module can possibly be viewed as a very high-level programming function, but since they exist as independent executables, these modules can very easily be chained using standard shell scripting, thus easily building complex applications. Examples of such modular systems are OS- SIM, SAGA, the Orfeo Toolbox, to a certain extent the GDAL and OGR command line tools and others. One of the oldest of these toolboxes is GRASS GIS.

The use of such existing, and highly optimized, bricks, which, in this context, can almost be considered as the elements of a "programming language" of their own, allows to quite rapidly construct a program for calculating desired spatial indicators, generally using some general-purpose programming language as glue. One can, therefore, concentrate on the higher-level logic and not on the concrete implementation. The modular structure, with each module being its own executive, makes the programs chaining them particularly robust, with failure in one module not necessarily meaning failure in the rest. It also provides an easy entry point to speeding up analyses through high-level parallel processing. In addition, the fact that all the components are open source allows to possibly adapt them if necessary, but, more importantly in a research context, allows to control how they work.

In this article I, thus, argue that using existing FOSS4G modular software al- lows the researcher in economic geography to easily elaborate efficient analyses on large datasets without having to focus on every aspect of programming. I present two examples of algorithms useful in economic geography (and other spatial disciplines) that were easy to code using a modular system such as GRASS GIS: the main example concerns the elaboration of new indicators for estimating regional exports based on firm micro-data. A second, small, example covers calculating a neighborhood matrix for raster objects. I begin by presenting the tools more in detail. I then explain each of the two use cases and the way they were implemented in GRASS GIS.

2 The tools

2.1 A modular GIS for efficient raster processing and projection handling

GRASS GIS (Neteler et al., 2012) is a generalist GIS software package that has been under constant development for over 30 years. Although it now offers very sophisticated vector handling, at its origin it was almost purely raster based and as such its raster routines have known a constant scrutiny and improvement over a long period of time. Because of this history, with much of the code written in times of much lower machine performances, many of these routines are particularly memory efficient and can thus easily handle large data sets.

GRASS GIS can be used as a straightforward desktop GIS, but its aforementioned modular structure invites the user to construct targeted applications through chaining of module calls. Two elements make the use of a GIS toolbox such as GRASS GIS particularly interesting for economic geography, especially when working in continuous space: i) handling location, space and projections and ii) existing (efficient) matrix/raster processing routines.

By definition economic geography studies the spatial aspect of economic production and development. In that context the notions of location and distance take up particular importance. Many theoretical economic geography studies use abstract representations of space, often only including a virtual distance in a virtual plane, without real localisation. Empirical studies on real firms in real space, however, need to be localised using the relevant coordinate systems: either non- projected latitude-longitude coordinates in degrees on a given ellipsoid or projected coordinates in meters on a plane defined by a specific projection system. Often different sources provide their data in different projection systems and data has to be reprojected from one to the other in order to be able to combine these different sources. Handling such projection issues is not easy, and errors can lead to wrong results, including erroneous distance measurements. As most of the FOSS4G ecosystem, GRASS GIS relies on a well-tested and constantly peer-reviewed open source system of projection handling, proj.4, and provides sophisticated options for projection management. Instead of having to worry about these issues on library level, the researcher can thus rely on these tools to provide the necessary processes with sufficient accuracy.

The second advantage of using GIS software when working in continuous space is the existence of routines handling raster data as a whole, including raster data that do not fit into the computer's memory. Using modules as bricks the researcher does not have to worry about the individual treatment of pixels, nor about processing data in chunks (row-tiles or tiles). GRASS GIS has particularly excellent memory handling, using something that can be considered as disk-based external memory. Implementing such routines from scratch is quite complicated and thus, using well-tested existing tools avoids reinventing the wheel.

2.2 The glue: Python

Although one can use any shell scripting language to elaborate processing chains by combining modules, GRASS GIS specifically proposes a simple python scripting API that allows to call the individual GRASS GIS modules from within a Python script. This allows combining the power of the GRASS GIS modules with the ease of an interpreted language such as Python for the necessary glue and house-keeping. Python is open-source, widely used and well-documented, notably in the geospatial community, and I will not present generalities here. Two Python libraries, however, were particularly useful in the context of this study: the GRASS Python scripting API and the multiprocessing module.

2.3 The GRASS Python scripting library

GRASS GIS provides two Python application programming interfaces (API) to its functionalities. A very simple Python scripting API which mainly aims at easing calls to GRASS GIS modules from Python code, and in addition adds a small series of specific convenience functions, and a more sophisticated, more low-level, API which provides access to internal C-function via ctypes, in addition to higher-level module access (Zambelli et al., 2013). Both allow to easily chaining GRASS GIS modules while profiting from a more sophisticated scripting language than those offered by standard command line shells.

In the examples here, I use the simple scripting API, essentially the run command(), read command(), parse command(), etc functions that allow to launch a GRASS GIS module and, if relevant, to capture its output. The core modules used in the central algorithms can all be called through run command(). Other smaller tasks, such as calculating the total sum all pixel values or collecting individual pixel values, can be done by calling read /parse command, loading the result into Python variables that can then be treated through Python tools.

For the scientific programmer this API is particularly easy to use as the translation from the command-line usage of the GRASS modules to the GRASS Python scripting API is very straightforward. A classic development flow would therefore start with interactive usage of GRASS GIS modules, possibly via the graphical user interface, to go on to the assembly of these modules into a simple shell script to then be finalised in a more complete python script (possibly a new GRASS module) using the API. Another alternative is to construct the script using the graphical modeller which provides the option to export the process as a Python script.

The accelerator: Simple parallel processing using Python's multiprocessing module

Researchers have to handle ever larger data sets. Very sophisticated environments and tools exist for high-performance computing dealing with really "big data". There are data sizes, however, which push the normal desktop computer to its limits, but for which the use of complex distributed file systems and cluster computing seem a bit overkill. Through its already mentioned memory handling, many of GRASS GIS' raster routines are mostly CPU-bound, not memory-bound. Using the increasing number of cores in standard desktop machines, can thus provide an easily accessible approach to significantly speed up processing time.

This is where the Python multiprocessing module comes in. It makes it easy to distribute the computation load across multiple nodes. It does not provide multithreading, but allows to launch parts of a Python script as multiple parallel processes. This allows to do high-level parallel processing, enough to gain significant calculation speed for medium to large datasets. The modular structure of toolboxes such as GRASS GIS, provide an excellent opportunity to divide tasks into separate processes. Thus, without any advanced programming knowledge and using just a few simple calls to the Python module's functions, one can easily spread the work-load amongst the increasing number of cores on the typical desktop machine, but also amongst the much larger number of cores available in university computing centers.

3 Example 1: Elaboration of new indicators for flexible-scale estimation of exports

3.1 Continuous space analysis in economic geography

Economic geography has long used, and still is using, aggregate data, mostly in administrative units. Such data is generally easily available and, through its aggregation, does not represent a heavy data load. Many indicators have been developed for such aggregated data and much useful information has been extracted from such indicators.

It has been known for a long time, however, that such aggregated data can cause a series of issues, non the least of which the modifiable areal unit problem (Briant et al., 2010; Gehlke and Biehl, 1934; Grasland and Madelin, 2006; Openshaw, 1983). Some voices plead for a renewed economic geography based on continuous-space analysis of data. Tobler and Grasland et al., as well as, and more specifically in the field of economic geography, Arbia, all propose the use of continuous space in order to transcend the issues raised by discrete boundaries, or the use of the flexibility of aggregation provided by point-located micro-data (Arbia, 2001a; Arbia et al., 2009, 2014a; Grasland et al., 2000; Tobler, 1979, 1989).

Micro-data forms the basis of such analyses and more and more analyses exist based on such data. As stated by Grasland and Madelin, "the access to individual data is of course the ideal situation, not because the individual level is in all cases the most appropriate one to observe or model a phenomenon, but mainly because it gives the choice to observe information at all possible levels and for all forms of spatial partitions" (Grasland and Madelin, 2006, p. 26). Examples of the use of such micro-data include birth, survival and growth analysis of firms (Arbia, 2001a; Arbia et al., 2014a, 2014b), clustering (Boix et al., 2011), the link between agglomeration and exports (Farole and Winkler, 2013; Koenig, 2009) and the heterogeneity literature (Bernard et al., 2012; Greenaway and Kneller, 2007; Melitz and Redding, 2015; Redding, 2011). However, dealing with such data creates new challenges for analysis, such as the development of new forms of analysis in continuous space, but also more practical challenge of dealing with geocoding the data and treating large amounts of data in an efficient manner. This is where the use of existing GIS tools can facilitate the work of the researcher.

3.2 Estimation of regional exports

Depending on the country, we have more or less precise data about international exports of firms, even at regional level. However, we know little about the pro- portion of the economic production that is exported to other regions in the same country.

Generally, estimates of the share of exported production are based on notions of concentration, most notably on location quotients, or derivatives thereof. Such approaches are often severely criticized (Isserman, 1980; Richardson, 1985), but are nevertheless subject to ongoing research and use (Billings and Johnson, 2012; Crawley et al., 2013; Strotebeck, 2010; Tian, 2013), because of their simplicity, but also because of a lack of credible and feasible alternatives. One of the biggest issues with concentration based analyses is the issue of the size of the space for which concentration is measured, often leading to the already mentioned modifiable areal unit problems (Arbia, 2001b; Tiebout, 1956).

In order to avoid such problems, in line with the literature analyzed above, I propose to work in continuous space using micro-data by adapting two existing indicators to the estimation of regional exports, based on existing work by Huff (1964) and by Marcon & Puech (2010). The details of these new indicators are explained elsewhere (Lennert, 2015), so I will only present them briefly and concentrate here on the practical implementation of their calculation.

3.3 Huff-like indicator

Huff (1964) defined his famous model in the context of market area and market share analysis. It relates the sales capacity of a given unit j (measured in number of employees) to the sales capacity of all sales units in the reference space, weighted according to their distance to the point i in space considered. The result gives the probability of resident at point i to consume at sales unit j:

$$P_{ij} = \frac{A_j / D_{ij}^{\gamma}}{\sum_j \left(A_j / D_{ij}^{\gamma} \right)}$$

I propose to use the same equation, but for all production units of a given sector, not only for retail, following the already mentioned basic idea that the more the production of a sector is concentrated in comparison to the consuming population the more this production is exported from the producing regions to other regions in the country. Consuming population can be employment, resident population or income, total production or any combination of these in order to combine intermediate and final consumption. In addition, this consuming population can be corrected by the relative trade balance of the sector at national level in order to take into account the amount of production that is exported or imported. In the following, however, I will concentrate on the simple version of the indicator as the mentioned additions do not change the fundamental logic.

The following algorithm represents the practical translation of the indicator into program flow:

- For each economic sector (NACE):
  - Extract all vector points representing firms of that NACE
  - Get γ of NACE (see below for details)
  - For each firm:
    - For each pixel:
      - Calculate distance to that firm
      - Calculate distance-weighted attraction to firm
  - For each pixel calculate sum of attractions to all firms
  - For each firm:
    - For each pixel:
      - Calculate probability of that pixel to buy products of firm
      - Calculate population of that pixel that buys products of firm

In the absence of empirical data, the choice of per-NACE γ coefficients is somewhat arbitrary, even in the original Huff model. For this exercise I simply linearly scaled the γ per NACE between 0 (no influence of distance) and 2 (very local sales area) depending on the distance between all points (pixels) in the total covered space and production units of that NACE. See the annex for the detailed algorithm.

In order to then analyze regional exports, several options are now open from there. One can:

- Calculate for each point in space the proportion of its population catered to by a specific area in space
- Calculate total population that is supplied by firms of given spatial units, possibly comparing that population to the resident population of these spatial units to deduce exports (thus going back to spatial aggregation, but after having calculated the spatial concentration in continuous space).

3.4 The modified Marcon-Puech indicator

Marcon & Puech (2010) propose a function, called the M function, that can be considered as a local location coefficient in varying radius reference spaces. The authors use it to measure the average spatial concentration of industries in a given sector. The function is defined as follows:

$$M_s(r) = \frac{\sum_i \left(\frac{\sum_{j \neq i} \left(c_s(i,j,r) \cdot w_j \right)}{\sum_{j \neq i} \left(c(i,j,r) \cdot w_j \right)} \right)}{\sum_i ((W_s - w_i)/(W - w_i))}$$

where i,j = firms, r = radius, s = sector, c = neighborhood function [0,1] within a given r, w = weight of firm (e.g. employment), W = total sum of weights.

One can either run this program with a different radius for each sector depend- ing on the general geography of its production units (similar to the γ coefficient in the Huff-model), or with a fixed radius for all sectors, possibly testing for the effect of distance by looping through a series of fixed radii around the production units and estimating trade balance or export employment rate by pixel for each radius.

In order to estimate exports, I rewrite their equation in line with Isserman's modification of the classic location quotient (Isserman, 1980):

$$X_{sr} = (L_{sr}/L_{sn} - L_r/L_n) \cdot L_{sn},$$

where $L_{sr} = \sum_j (c_s(i,j,r) \cdot w_j)$ and $L_{sn} = W_{s\cdot}$.

Again, one can refine this equation with the introduction of different measures of consumption and by taking into account the sectors relative international trade balance, but I will not do that here.

Translating this equation into algorithmic logic gives the following program flow:

- For each NACE:
  - Create a raster map with sum of employment per 1km pixel
  - Determine radius of reference space
  - For each pixel (and the respective reference space around the pixel):
    - Calculate total employment in that NACE within the reference space
    - Calculate total employment of all NACE combined within the reference space
    - Estimate employment trade balance within the reference space based on location quotient (in comparison to national rates)
    - Possibly: If trade balance is > 0: Calculate total export employ- ment of the pixel by multiplying export employment rate by the total employment in that pixel

3.5 Translating these equations into GRASS GIS "language"

This section presents the practical implementation of the above theoretical developments in GRASS GIS. For each indicator, I present the modules used in the core of the analysis. I do not present the complete program flow, including general input output handling and intermediate storage of results, but only the part which implements the fundamental elements. In order to ease readability, I also do not present the code in its Python API form, but as pure GRASS module calls. The entire source code of each example of these export estimation indicators as well as the code for a more sophisticated evolution of the Huff-like indicator into a doubly-constrained spatial interaction model can be found at https://github.com/mlennert/RegionalTradeModel/.

3.6 Huff-like indicator

For the Huff-like indicator, the first loop across all pixels to calculate the distance- weighted attraction to a given production unit (i.e. the nominator in equation 1) was easily calculated with fundamentally two module calls for each production unit:

```
r.grow.distance MapOfFirm output=distance
r.mapcalc "PerPixelAttractiveness = NbrEmployees / (distance^gamma)"
```

Here, `distance` is a map with for each pixel the distance to the production unit. `r.mapcalc` is a map calculator that applies the given formula pixel by pixel, with `NbrEmployees` and `gamma` available respectively for each production unit or for each NACE code.

In order to calculate the share of the population of a pixel that is catered to by a specific production unit, again only very few module calls are necessary:

```
r.series file=ListOfAttractivenessMaps method=sum output=TotalSum
For each production unit:
 r.mapcalc "PerPixelShare = PerPixelAttractiveness / TotalSum"
 r.mapcalc "PerPixelConsumingPopulation = PerPixelShare *
   PerPixelPopulation"
```

`r.series` creates a map with pixel-by-pixel aggregate statistics of all input maps which are given here through a text file `ListOfAttractivenessMaps` that contains a list of all relevant map names.

The rest of the code is mostly just logistics (extracting the point of location of each firm from a vector map, transforming to raster, correcting the distance measure for the pixel in which the firm is located, compiling the text file with the list of input maps, etc) or housekeeping (taking stock of the existing maps, erasing temporary maps, etc).

3.7 The modified Marcon-Puech indicator

As for the Huff-like indicator, the main core of the calculation of the Marcon- Puech-inspired indicator consists of only three module calls, based on existing per-NACE maps of total employment per pixel:

```
r.neighbors input=EmploymentPerPixelPerNACE method=sum nsize=radius \
  output=NaceEmploymentNeighborhood

r.neighbors input=TotalEmploymentPerPixel method=sum nsize=radius \
  output=TotalEmploymentNeighborhood

r.mapcalc "ExportEmploymentNeighborhood = \
  (NaceEmploymentNeighborhood / NaceEmploymentReference - \
  TotalEmploymentNeighborhood/TotalEmploymentReference) \
  NaceEmploymentReference"
```

The module `r.neighbors` provides a map with moving window stats for each pixel in the radius `nsize` expressed in number of pixels.

3.8 Parallelization of the processes

For both of the above routines, where each calculation has to be done NACE by NACE (and sometimes even firm by firm), parallel processing can significantly increase processing speed, thus allowing to upscale the analyses to even larger data sets. Parallel processes fetch the next non-treated NACE code in an input queue listing of all sectors to be treated, launch the necessary processing and either add alphanumeric results to a result queue, or just create new output maps. The fact that each module call can be launched as a separate process, as each GRASS GIS module is an autonomous executable, this very high-level approach is straighforward to implement, by adding just a few lines of code. As much documentation exists, I will not go further into details here. The code will be in the complementary material accompanying the final version of the paper.

4 Example 2: Calculating neighborhood matrices of raster objects

The notion of proximity is central in spatial analysis and very much so in economic geography. Tobler's first law of geography states that "everything is related to everything else, but near things are more related than distant things" (Tobler, 1970). Many models in geography are based upon this law and many tools exist to evaluate the importance of proximity, notably spatial autocorrelation (Geary, 1954; Moran, 1950). In many of these analyses, the notion of neighborhood plays a particularly important role. Tools already exist in FOSS4G environments for the elaboration of neighborhood matrices. In GRASS GIS, the addon module v.neighborhoodmatrix does just that for vector polygons, profiting of the topological nature of the internal vector format (Lennert and GRASS GIS Development Team, 2014). R's spdep package provides flexible options for analyzing neighborhood relations (Bivand et al., 2013; Bivand and Piras, 2015). However, none of these seem to take into account the situation where objects (i.e. clumps of pixels) are represented in raster maps. Such objects can obviously be vectorized and submitted to vector-based algorithms, but sometimes going through such an extra step can be computation- ally intensive. In the context of the elaboration of tools for object-based image analysis for GRASS GIS, the need thus arose for the elaboration of a tool that can provide a neighborhood matrix of raster objects. In this, very brief, section I show how this was implemented in the form of a new publicly available addon module r.neighborhoodmatrix (Lennert and GRASS GIS Development Team, 2016).

Raster objects are defined by clumps of adjacent pixels that have the same pixel value. This value is considered the object id. The algorithm to determine a neighborhood matrix based on such maps is quite simple:

- For each direction (4, 8 or even more):
  - For each pixel i:
    - Check if neighboring pixel value V_j ! = V_i
    - If true: add i, j to the list of neighbors

- If false: ignore

The practical implementation of this algorithm in GRASS GIS is again also very straightforward and consists in basically two module calls for each direction:

```
r.mapcalc "NeighborMap = if( \
  OriginalMap[NeighborhoodModifier] != OriginalMap, \
  OriginalMap[NeighborhoodModifier], \ null())"
r.stats -1n input=OriginalMap,NeighborMap
```

The `r.mapcalc` call has to be read as follows: if(expression, then, else). For a 4-direction analysis, `NeighborhoodModifier` is one of [0,-1], [0,1], [1,0], or [-1,0], representing respectively the left, right, upper and lower neighbor. Each pixel in `NeighborMap` thus contains the id of the neighbor if it is different, and a null value otherwise.

With the '1' flag set r.stats exports pixel values one by one and if more than one map is given as input (as is the case here) each output row contains the values of the equivalent pixels of all input maps. The 'n' flag instructs it to ignore null values, thus only taking into account those pixels that actually are on the border between two objects. The GRASS Python scripting API's `read_command()` function is used to capture the output of `r.stats` into a Python list which is filled direction by direction. As the result of these two module calls is pixel by pixel, and not object by object, as well as direction by direction, the list contains multiple entries of neighborhood pairs. A last step thus consists in uniquifying the list of neighborhood relations. This can be done by a simple Python call:

```
unique_neighbors = list(set(neighbors))
```

The rest of the code just handles general input and output. This second small example shows, again, how the use of a modular GIS tool- box such as GRASS GIS, allows to code algorithms in a very high-level language which takes away all of the hassle of handling the spatial data in detail. In order to accelerate the process, it is obviously possible to divide the per-direction r.mapcalc and r.stats calls amongst several computational cores. Again, this can be done very easily using Python's multiprocessing module.

5 Conclusion

The main aim of this paper was to show that using the modules of a modular GIS toolbox such as GRASS GIS (or others such as SAGA or Orfeo Toolbox for image processing) as bricks allows to very easily and quickly construct more elaborate spatial analyses. As mentioned, the modules can be envisioned as elements of a particular programming language. In combination with a more general-purpose language such as Python, very sophisticated applications can be built quite easily.

The main difficulty, in my experience, is to change the paradigm of thinking from an individual data point for data point (in this case pixel for pixel) perspective to a map perspective, trusting the GIS modules to handle the looping across all of the data. Once this change of perspective is realized, what used to seem complicated algorithms suddenly become quite easy to implement.

GRASS GIS' excellent memory handling in its raster modules, fruit of its long history and some significant revisions along the way, brings the additional added value of freeing the researcher of having to deal with the memory issues raised by large data sets. GRASS GIS modules can process gigabyte-heavy datasets while only using a few hundred megabytes of memory.

For CPU-bound applications, the combination of the modular approach, potentially treating each call to a GIS function as a separate process, allows the use of very simple parallelization strategies, such as Python's multiprocessing module. This approach cannot always rival with specifically programmed, targeted multi-threaded routines, but it can be a very good compromise between the need for upscaling and the limited time and energy available for programming.

Thus, as soon as the analysis the researcher aims to implement goes beyond the standard set of tools a normal spatial analysis or GIS package provides, combining modular elements makes application development easier, and possibly more efficient by using simple multi-processing. Obviously, one is not limited to only one set of modules, but can combine, in one single application, calls to the modules of different toolboxes.

This is also where the strength of FOSS4G comes in: although many proprietary tools also provide such modular access to functionality, each additional toolbox adds to the overall licencing costs and it is not always easy to combine tools from different vendors. The free and open nature of FOSS4G projects, combined with the communication between developers, such as at FOSS4G conferences, creates an ecosystem where teams collaborate and thus ease the combination of different tools, without extra costs. In addition, the open source nature also provides the possibility to review the code, down to the very routines provided by the toolboxes, something that has gained in importance as efforts increase towards openness and reproductibility of research. As others have argued, peer review of code is just as important as peer review of methods and results (Barnes, 2010; Ince et al., 2012; Nosek et al., 2015).

6 References

Anselin, L., 2000. Part 2 The Link between GIS and spatial analysis. Journal of Geographical Systems 2, 11–15. doi:10.1007/s101090050023

Arbia, G., 2001a. Modelling the geography of economic activities on a continuous space. Papers in Regional Science 80, 411–424. doi:10.1111/j.1435-5597.2001.tb01211.x

Arbia, G., 2001b. The role of spatial effects in the empirical analysis of regional concentration. J Geograph Syst 3, 271–281. doi:10.1007/PL00011480

Arbia, G., Cella, P., Espa, G., Giuliani, D., 2014a. A micro spatial analysis of firm demography: the case of food stores in the area of Trento (Italy). Empirical Economics 1–15. doi:10.1007/s00181-014-0834-6

Arbia, G., Copetti, M., Diggle, P., 2009. Modelling Individual Behaviour of Firms in the Study of Spatial Concentration, in: Fratesi, D.U., Senn, P.L. (Eds.), Growth and Innovation of Competitive Regions, Advances in Spatial Science. Springer Berlin Heidelberg, pp. 297–327.

Arbia, G., Espa, G., Giuliani, D., Dickson, M.M., 2014b. Spatio-temporal clustering in the pharmaceutical and medical device manufacturing industry: A geographical micro-level analysis. Regional Science and Urban Economics 49, 298–304. doi:10.1016/j.regsciurbeco.2014.10.001

Barnes, N., 2010. Publish your computer code: it is good enough. Nature News 467, 753–753. doi:10.1038/467753a

Bernard, A.B., Jensen, J.B., Redding, S.J., Schott, P.K., 2012. The Empirics of Firm Heterogeneity and International Trade. Annual Review of Economics 4, 283–313. doi:10.1146/annurev-economics-080511-110928

Billings, S.B., Johnson, E.B., 2012. The location quotient as an estimator of industrial concentration. Regional Science and Urban Economics 42, 642–647. doi:10.1016/j.regsciurbeco.2012.03.003

Bivand, R., Hauke, J., Kossowski, T., 2013. Computing the Jacobian in Gaussian Spatial Autoregressive Models: An Illustrated Comparison of Available Methods: Computing the Jacobian in Spatial Autoregressive Models. Geographical Analysis 45, 150–179. doi:10.1111/gean.12008

Bivand, R., Piras, G., 2015. Comparing Implementations of Estimation Methods for Spatial Econometrics. Journal of Statistical Software 63. doi:10.18637/jss.v063.i18

Boix, R., Lazzeretti, L., Oliver, J.L.H. s, Molina, B.D.M., Ruiz, B.T., 2011. Creative clusters in Europe: a microdata approach (ERSA conference paper No. ersa11p471). European Regional Science Association.

Briant, A., Combes, P.-P., Lafourcade, M., 2010. Dots to boxes: Do the size and shape of spatial units jeopardize economic geography estimations? Journal of Urban Economics 67, 287–302. doi:10.1016/j.jue.2009.09.014

Crawley, A., Beynon, M., Munday, M., 2013. Making Location Quotients More Relevant as a Policy Aid in Regional Spatial Analysis. Urban Stud 50, 1854–1869. doi:10.1177/0042098012466601

Farole, T., Winkler, D., 2013. Firm location and the determinants of exporting in low- and middle-income countries. J Econ Geogr lbs060. doi:10.1093/jeg/lbs060

Fotheringham, A.S., 2000. Context-dependent spatial analysis: A role for GIS? Journal of Geographical Systems 2, 71–76. doi:10.1007/s101090050032

Geary, R.C., 1954. The Contiguity Ratio and Statistical Mapping. The Incorporated Statistician 5, 115. doi:10.2307/2986645

Gehlke, C.E., Biehl, K., 1934. Certain Effects of Grouping Upon the Size of the Correlation Coefficient in Census Tract Material. Journal of the American Statistical Association 29, 169. doi:10.2307/2277827

Grasland, C., Madelin, M., 2006. The Modifiable Area Unit Problem (Final Report No. 3.4.3). ESPON.

Grasland, C., Mathian, H. 'el\`ene, Vincent, J.-M., 2000. Multiscalar analysis and map generalisation of discrete social phenomena: Statistical problems and political consequences. Statistical Journal of the United Nations Economic Commission for Europe 17, 157–188.

Greenaway, D., Kneller, R., 2007. Firm heterogeneity, exporting and foreign direct investment*. The Economic Journal 117, F134–F161. doi:10.1111/j.1468-0297.2007.02018.x

Griffith, D.A., Chun, Y., 2018. GIS and Spatial Statistics/Econometrics: An Overview, in: Comprehensive Geographic Information Systems. Elsevier, pp. 1–26. doi:10.1016/B978-0-12-409548-9.09680-9

Huff, D.L., 1964. Defining and Estimating a Trading Area. Journal of Marketing 28, 34–38. doi:10.2307/1249154

Ince, D.C., Hatton, L., Graham-Cumming, J., 2012. The case for open computer programs. Nature 482, 485–488. doi:10.1038/nature10836

Isserman, A.M., 1980. Estimating Export Activity in a Regional Economy: A Theoretical and Empirical Analysis of Alternative Methods. International Regional Science Review 5, 155–184. doi:10.1177/016001768000500204

Koenig, P., 2009. Agglomeration and the export decisions of French firms. Journal of Urban Economics 66, 186–195. doi:10.1016/j.jue.2009.07.002

Lennert, M., 2015. The Use of Exhaustive Micro-Data Firm Databases for Economic Geography: The Issues of Geocoding and Usability in the Case of the Amadeus Database. ISPRS International Journal of Geo-Information 4, 62–86. doi:10.3390/ijgi4010062

Lennert, M., GRASS GIS Development Team, 2016. r.neighborhoodmatrix. Open Source Geospatial Foundation, USA.

Lennert, M., GRASS GIS Development Team, 2014. v.neighborhoodmatrix. Open Source Geospatial Foundation, USA.

Marcon, E., Puech, F., 2010. Measures of the geographic concentration of industries: improving distance-based methods. J Econ Geogr 10, 745–762. doi:10.1093/jeg/lbp056

Melitz, M.J., Redding, S.J., 2015. Chapter 1 - Heterogeneous Firms and Trade, in: Elhanan Helpman, K.R. and G.G. (Ed.), Handbook of International Economics, Handbook of International Economics. Elsevier, pp. 1–54.

Moran, P.A.P., 1950. Notes on Continuous Stochastic Phenomena. Biometrika 37, 17–23. doi:10.2307/2332142

Neteler, M., Bowman, M.H., Landa, M., Metz, M., 2012. GRASS GIS: A multi-purpose open source GIS. Environmental Modelling & Software 31, 124–130. doi:10.1016/j.envsoft.2011.11.014

Nosek, B.A., Alter, G., Banks, G.C., Borsboom, D., Bowman, S.D., Breckler, S.J., Buck, S., Chambers, C.D., Chin, G., Christensen, G., Contestabile, M., Dafoe, A., Eich, E., Freese, J., Glennerster, R., Goroff, D., Green, D.P., Hesse, B., Humphreys, M., Ishiyama, J., Karlan, D., Kraut, A., Lupia, A., Mabry, P., Madon, T., Malhotra, N., Mayo-Wilson, E., McNutt, M., Miguel, E., Paluck, E.L., Simonsohn, U., Soderberg, C., Spellman, B.A., Turitto, J., VandenBos, G., Vazire, S., Wagenmakers, E.J., Wilson, R., Yarkoni, T., 2015. Promoting an open research culture. Science 348, 1422–1425. doi:10.1126/science.aab2374

Openshaw, S., 1983. The modifiable areal unit problem. Geo Books, Norwick [Norfolk].

Redding, S.J., 2011. Theories of Heterogeneous Firms and Trade. Annual Review of Economics 3, 77–105. doi:10.1146/annurev-economics-111809-125118

Richardson, H.W., 1985. INPUT-OUTPUT AND ECONOMIC BASE MULTIPLIERS: LOOKING BACKWARD AND FORWARD*. Journal of Regional Science 25, 607–661. doi:10.1111/j.1467-9787.1985.tb00325.x

Strotebeck, F., 2010. The Location Quotient – Assembly and application of methodological enhancements [WWW Document]. URL http://mpra.ub.uni-muenchen.de/47988/ (accessed 10.21.14).

Taupier, R., Willis, C., 1994. Geographic Information Systems and Applied Economics: An Initial Discussion of Potential Applications and Contributions. Agricultural and Resource Economics Review 23, 140–149.

Tian, Z., 2013. Measuring Agglomeration Using the Standardized Location Quotient with a Bootstrap Method. Journal of Regional Analysis & Policy 43, 186–197.

Tiebout, C., 1956. Exports and Regional Economic Growth. Journal of Political Economy 64, 160–164. doi:10.1086/257771

Tobler, W.R., 1989. Frame independent spatial analysis, in: Goodchild, M.F., Gopal, S. (Eds.), The Accuracy of Spatial Databases. CRC Press, London, Great Britain, pp. 115–122.

Tobler, W.R., 1979. Smooth Pycnophylactic Interpolation for Geographical Regions. Journal of the American Statistical Association 74, 519–530. doi:10.2307/2286968

Tobler, W.R., 1970. A Computer Movie Simulating Urban Growth in the Detroit Region. Economic Geography 46, 234. doi:10.2307/143141

Zambelli, P., Gebbert, S., Ciolli, M., 2013. Pygrass: An Object Oriented Python Application Programming Interface (API) for Geographic Resources Analysis Support System (GRASS) Geographic Information System (GIS). ISPRS International Journal of Geo-Information 2, 201–219. doi:10.3390/ijgi2010201

Annexe

Calculation of the gamma coefficient in the Huff model

- For each NACE:
    - Calculate distance from each pixel in the total space to the closest production unit in that NACE
    - Calculate summary statistics (e.g. 90th percentile) of these distances
- Chose one of the summary statistics
- Determine the per-NACE γ as:
    - 0 if summary statistics of NACE is above a threshold max value of distance (i.e. distance does not really matter)
    - 2 if summary statistics of NACE is below a certain threshold min value (i.e. very local distribution of goods)
    - else: scale linearly between 0 and 2 according to the position of the summary statistic between the max and min threshold values

Free and Open Source Software for Geospatial (FOSS4G) Conference Proceedings

Volume 16 *Bonn, Germany* Article 4

2016

Involving communities in environmental protection by Community Information Systems: the case study of "La Cuicadora"

Domenico Vito
Politecnico di Milano, Italia

Recommended Citation

Vito, Domenico (2016) "Involving communities in environmental protection by Community Information Systems: the case study of "La Cuicadora"," *Free and Open Source Software for Geospatial (FOSS4G) Conference Proceedings*: Vol. 16 , Article 4.
DOI: https://doi.org/10.7275/R5KW5D7S
Available at: https://scholarworks.umass.edu/foss4g/vol16/iss1/4

Involving communities in environmental protection by Community Information Systems: the case study of "La Cuicadora"

Domenico Vito

Politecnico di Milano – domenico.vito@polimi.it

KEYWORDS: Participatory GIS, Participatory Sensing, Crowdmapping, Sustainable Development, Environmental Protection

ABSTRACT:

The involvement of the communities is a key-strategy to enable fast responses both case of emergency and environmental protection.

On this goal geospatial system could allow to collect information directly from the citizens. More specifically they can act on "feedback-loops" between "communities", "decision-makers" and "environment" catalysing participation and the perception of local knowledge. The work will present the case study of "la Cuicadora" project, a Ushaidi based crowdmapping system provides an information sharing network to assist Peruvian indigenous communities in better protecting themselves from contaminated water sources.

The analysis offers the chance to understand the concept of Community Information Systems (CIS) and its role in the involvement of communities in environmental protection.

1 Introduction

Promoting participation through community development projects and local decentralization has become a central tenet in environmental protection.

Although the participatory approach has certainly not a narrow focus, it surely concerns the involvement of the population in the planning and implementation of community policies.

Such "holistic" form of people's participation are certainly required for area-based operations which affect all inhabitants like environmental protection, soil and water conservation, provision of physical, economic and social infrastructures and irrigation, sanitation and health schemes.

The necessity of public participation for successful environmental policy has been recognized for some years at both the international and the national level.

As implemented in Agenda 21 one of the most important focus for get environmental policies more focused on communities is the participatory approach and public participation, including participation by specific sections of the population, e.g. women, children, indigenous peoples and farmers (Eden, 1996).

Considering the capillarity and the community contribution to environmental problem solving, the diffuse involvement of society within environmental issues could represent a key factor for the success of those actions.

Successful environmental policy has therefore been linked to the notion of 'concerned citizens', coupling individual activism to institutional action (Kingston et al. 2000).

The 'concerned citizen' is strictly related to the way the information about the specific topic comes from the top to the bottom (Macnaghten et al. 1995), and how the mechanism of involvement about the specific problem reaches both policy maker and the stakeholders.

Such mechanism nowadays could be strongly mediated by geographical information system, and in particular the use of Web represent a bidirectional way to realize participation, in the sense that they allow to gain information's, notices, ideas directly from the citizens. By the involvement of communities, Geospatial Information Systems through the Internet, can be widened into Community Information Systems (Fraternali et al. 2000), that can be used by both institutional political actors, by civil societal actors, and even by single individuals enhancing bidirectional mechanisms of information gathering.

This work proposes the case study of "la Cuicadora" project, a crowd mapping system provides an information sharing network to assist Peruvian indigenous communities in better protecting themselves from contaminated water sources. La "Cuicadora" represents an example of Community Information System, that offers a system for fast response to environmental problems.

At first, the basic concept of sustainable development and participation, and their relationship will be introduced. They represent the theoretical basis and the keys of understanding for all following discussion. Subsequently a focus on participatory mapping and sensing will be proposed to finally present the case study of "La Cuicadora". The case study will be the chance to identify the role of CIS in environmental protection.

1.1 Participation and relationship with lands

More than 2000 years ago, the Greek philosopher Aristotle defined citizens as all who share in the civic life of ruling and being ruled in turn, and considered as good citizens who brings the knowledge and the skills both to rule and to be ruled (Gigler & Savita 2014). This definition of citizenship is strongly connected to the concept of participation.

There is a wide range of definitions and interpretations of participation.

In more descriptive terms it stands for; "the involvement of a significant number of persons in the decision-making process which regards their development" (Mansuri & Vijayendra 2013).

Thus participation has a strong relationship with the concept of development and furthermore with the relationship of a community with its inhabited land.

Classical approaches identifies the sustainable development as the intersections among economic, social and environmental dimension (Figure 1.a).

This model implies that some parts of the economics system are independent from the social system. and on the other side that some social structure could exist without a natural system which supplies natural resource.

To have an exhaustive vision for an effective strategy of sustainable development; D'Alisa (2007) propose a concentric framework where a fourth dimension is added, that is participative democracy (Figure 1.b).

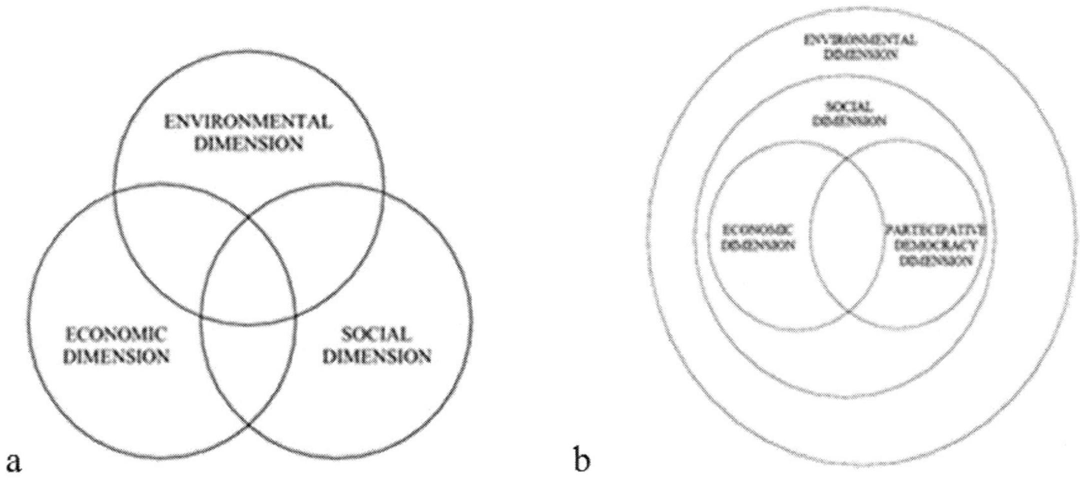

Figure1: a) classical approach b) concentric dimensions of sustainable development

As can be deduced from the graphical representation, participation and sustainability have a strict relationship of inter-dependency. This relationship has not linear fashion of cause-effect, but stands in a circular shape that traduce the complex link between a inhabitant community and its territory.

1.2 The role of local knowledge

Local knowledge is a key element for sustainable development and environmental management.

Standing to Minang and McCall (2006), it has also the characteristics of an informative system, in fact:

- it is a system of spatial information that is developed by the close relationship between the local population, its land and its natural resources
- Members of the community have a pool of experiential data of different categories according to their age, and social status
- it is an original knowledge of the local community.
- the system is a "pseudo" in the sense that consists in the classification of facilities and employs certain methodologies (e.g. Oral transmission)
- it is a holistic system in that it uses for decision different areas of knowledge

For this reasons is important to include local knowledge, as significant source of information when a specific action is done within a territory.

However as informative system, local knowledge has obviously some drawbacks as it has no facilities prediction, it has some gaps in transmission and finally it has little quantization.

1.3 Spatial Data Infrastructure to improve local knowledge involvement

Spatial Data Infrastructures (SDIs) are usually intended as the framework of technologies, policies, and institutional arrangements that together facilitate the creation, exchange, and use of geospatial data and related information resources across an information-sharing community .

Such a framework includes all the technologies like Geographical Information Systems (GIS), geoportals and diffuse georeferenced infrastructures that enable the sharing of geospatial information within an organization or more broadly for use a national, regional or global level.

In particular, if we consider a delimited area framed in a specific space and time, SDI can catalyze "feedback-loops" between the main actors involved in the processes related to that area (Figure 2).

This actor can be defined as communities, decision-makers and environment. Participation, empowerment and sustainability lies within the feedback loops as driving forces of circular relationships as explained above.

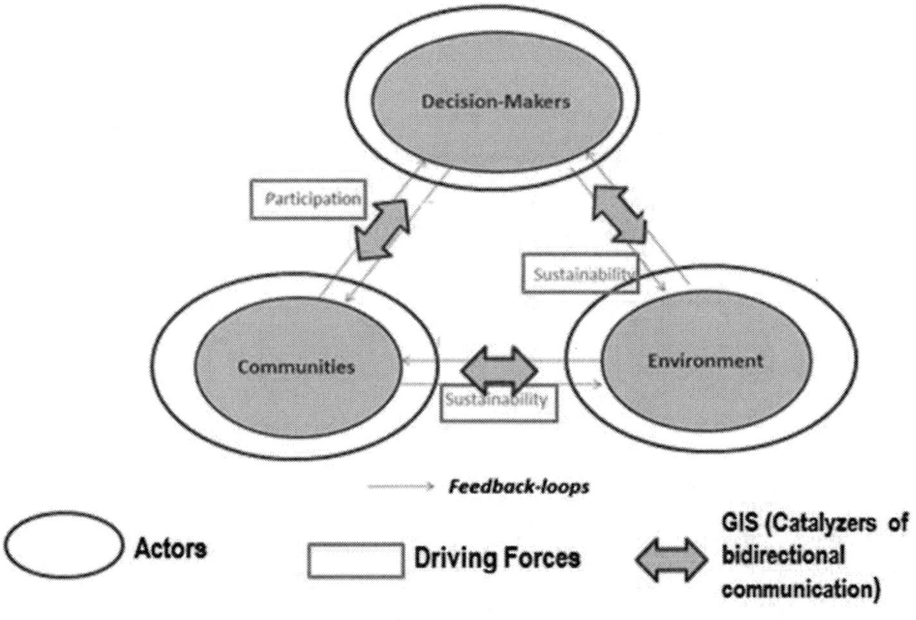

Figure 2: Informational Flux between institution, citizens, providers "feedback-loops" between "communities", "decision-makers" and "environment" catalyzing participation (Gigler & Savita 2014).

Central to the information flows catalysis, is the question of how to close the "feedback loop" between the involved actors. One approach is to "put citizens in the circuit"(Fraternali et al. 2000).

As citizens will be considered inside the loop, local knowledge could be part of process, and thus thanks to GIS, it comes to be organized in a spatial and a temporal scale. Public Participatory GIS (PPGIS), human computation and participatory sensing are some approaches that realize the involvement of citizens in the loop.

2 Participatory Mapping

In its broadest sense, participatory mapping means the creation of maps by local communities – often with the involvement of supporting organizations including governments, NGOs or other actors engaged in development or land-related planning (Lienert 2009).

Participatory Mapping is a well-known practice that can contribute to building community cohesion, help to engage participants to be involved in resource and land-related decision-making, raising awareness about pressing land-related issues. Participatory approaches in spatial analysis are rooted in the implementation of Participatory Rural Appraisal (PRA) methods during the 198Os. Their success shows the fast growth of people's participation (Oakley 1991). During the 1990s, Participatory Rural Appraisal (PRA) and Geographical Information Systems (GIS) came together for delivering Public Participatory Geographical Information Systems (PGIS).

The term public participatory GIS (PPGIS), originated during two meetings of the National Center for Geographic Information and Analysis (NCGIA) (NCGIA 1996a, 1996b) is aimed at defining the use of GIS technologies to expand public involvement in the policy to promote the objectives of local communities and grassroots organizations (Sieber 2006).

Geographical data and community maps may support the participatory approach during the data integration process by the ability to translate the gathered data into an explicit map location, as geospatial and time-series. However the participatory approach become effective only on certain conditions

Basically, maps represent an intermediary output of the long-term process and need to be integrated into the networking and communication initiatives.

The maps produced and the spatial analyses represent important steps in the process.

2.1 Human Computation and Participatory sensing

Participatory sensing can be defined as the use of mobile devices to form sensor networks that enable interactive and participatory to public and private users to collect, analyze and share local knowledge.

The participatory sensing implies the exchange of data through the collaboration of the users of mobile devices, thus obtaining a direct "feedback mechanism" with individuals.

Mobile devices in facts are increasingly capable of capturing, classifying and transmitting image, acoustic, location and other data, interactively or autonomously. Given the right architecture, they could act as sensor nodes and location-aware data collection instruments.

Including consumer devices as a fundamental building block of the sensing system implies that the human owners of these devices play an important role in the resulting system architecture.

The application of participatory sensing realizes the implementation of the paradigm of "human computation" (Quinn & Bederson 2012).

In the Human Computation Paradigm (Law & Von Ahn 2011) the interaction among users is harnessed to help in the cooperative solution of tasks.

Human Computation (HC) is usually applied in business, entertainment and science, but can also useful in the management of environmental resources, which are by definition shared and distributed and demand new approaches to their management, based on an increased consciousness of mankind's collective responsibilitie.

This approach connects with the concept of "concerned citizen", representing a way to open the informative stream between institutions and civic engagement.

Crowdsourcing and crowdmapping can be considered cases of HC.

Human computation opens the way to a more open and collaborative governance and environmental management..

2.2 Crowdmapping technologies: Ushaidi

Ushahidi Inc. is a non-profit software company that develops free and open-source software (LGPL) for information collection, visualization, and interactive mapping. Ushahidi created a well-known web-platform in the aftermath of Kenya's disputed 2007 presidential election (see 2007–2008 Kenyan crisis) that collected eyewitness reports of violence reported by email and text message and placed them on a Google Maps map (Okollo 2009).

The Ushahidi platform includes crowdmapping tools, which allows location data points sent by SMS message to be mapped, and Swiftriver, a data filtering system to control information flow in crisis situations so it can be effectively and rapidly employed. Ushahidi's website includes instruction videos, a wikispace, datasets, other literature and a blog. Ushaidi technologies constitute a diffuse crowdbased SDI: in a broader sense the constitute a of Community Information Systems (CIS).

2.3 Community Information System

A Community Information System (CIS) is normally an spatial information aggregator that's able to interact with user-citizens by means that are familiar to them and that are available and used in their everyday life (Fraternali et al. 2000),

It generally aims at organizing the citizen participation activities in the areas of risk and resource management as the case of "La Cuicadora"

3 Case Study: La Cuicadora

The project, called The Cuidadora, "the guardian" was born on initiative of Jonathan Rupire (Rupire 2015). It provides network shared information to help indigenous communities to better protect themselves from the dangers of contaminated water. It also report social-environmental conflicts and corruption.

The "Cuicadora" is an interactive crowdmapping platform for the collaborative georeferencing and dissemination of real-time alerts. The "Cuicadora" stems from the problem of contamination of water sources water due to the strong presence of sites for the extraction of oil. Using a distribution Ushahidi (translated into indigenous Shipibo), the information provided by affected communities are reported, mapped and then distributed to national organizations and media, as well as the communities themselves. The website is hosted on a server collective Ourproject.org.

It consists of a main page that displays a georeferenced map in the foreground.

Figure 3: Home page of "La Cuicadora" (Source: http://cuidadora.ourproject.org/)

The main menu is put over the map and allows to access to the various sections. The alert can be sent by registered user that can also receive alerts via email and SMS when the access to internet is unavailable. The reported events are geo-referenced and marked on the map. They are also grouped according to specific categories that can be expanded by users' suggestion. Section "recibe alertas" will also organize a period digest of all the repot displayed through the title, the description, time and photo within the specific category. The "Cuicadora" system is also provided by a Twitter account where the alert "tweets" is reported by specific signaling hashtags.

4 Conclusions

The case study of "la Cuicadora"; represents an example of "community information system" based alert system. The CIS infrastructure is able to spread fast response and equally fast propagations over all the places where it insists.

These two features are essential for an alert system, and became more effective as the number of user increases. As discussed in this work, "feedback-loops" could be strongly mediated by spatial information systems and thus CIS.

The way by which it is realized is related to the capability of information systems to catalyze bidirectional communications.

Bidirectional communication contributes to strength two mechanisms related to a community within its environment that are:

- the information-decision mechanism
- the action-reaction mechanism.

If we consider the first mechanism, it's possible to argue that the use of diffuse spatial information technologies enhances sustainability because they help to find the best fitted solution to a specific problem considering the boundary conditions related to environment and community needs. By diffuse spatial information the 'boundary conditions' of territorial problems are better known, due to the strengthening of the information-decision mechanism. This is more true if communities are involved furnishing the local knowledge of the a specific environment. The reinforcement of the action-reaction mechanism indeed and the catalysis of the feedback-loops between environment-communities and decision makers, could reflect on faster and better responses to environmental hazards.

In other words this means an improvement in resilience. So, it should be said that the use of a CIS help to foster the sustainable development and the resilience of the involved community, increasing furthermore the awareness of the population on environmental problems as result of citizen involvement in the process of territorial information gathering.

References

Eden, S. (1996), 'Public participation in environmental policy: considering scientific, counter-scientific and non-scientific contributions' Public Understand. Sci. 5, 183–204.

Kingston, R., Carver, S., Evans, A. & Turton, I. (2000), 'Web-based public participation geographical information systems: an aid to local environmental decision-making Computers', Environment and Urban Systems 24, 109-125.

Fraternali, P. , Castelletti, A., Soncini-Sessa, R. , Vaca Ruiz, C. & Rizzoli.,A.E. (2012), 'Putting humans in the loop: Social computing for Water Resources Management', Environ Model Softwares 37, 68–77.

Gigler, B.S. & Savita, B. (2014),'Closing the Feedback Loop: Can Technology Bridge the Accountability Gap?' Directions in Development, World Bank, Washington DC, 1-50.

Mansuri, G. & Vijayendra R. (2013), 'Localizing Development: Does Participation Work?', World Bank, Washington DC, 1-20.

Alisa, G. D., (2007), ' Dimensions of sustainable development : a proposal of systematization of sustainable approaches'. Department of Economics, Mathematics and Statistics, University of Foggia, Italy, Quaderno n. 9/2007.

Minang, A. & McCall, M.K. (2006), 'Participatory GIS and local knowledge enhancement for community carbon forestry', Participatory learning and actions, 85-92.

Lienert, J. (Last Checked 2012), 'Participatory Mapping for Decision Making', URL: http://www.sswm.info/content/participatory-mapping-decision-making

Oakley P. & al (1991), 'Projects with people: the practice of participation in rural development', The Journal of Agricultural Science 117, 385-385.

Sieber R.(2006), 'Public participation geographic information systems: A literature review and framework', AAG Ann, 491-507.

Quinn, A.J. & Bederson, B.B., 'Human computation: a survey and taxonomy of a growing field'. in: 'Proceedings of the 29th International Conference on Human Factors in Computing Systems'(Vancouver, 7 May 2011), 1403-1412.

Van Heck, B.(2003), 'Participatory Development: Guidelines on. Beneficiary Participation in Agricultural and Rural Development Rural Institutions and Participation, Rural Institutions and Participation Service', Service Rural Development Division FAO, 1-77.

Okolloh, O. (2009), 'Ushahidi, or "testimony": Web 2.0 tools for crowdsourcing crisis information'. Participatory Learning and Action, 59(January), 65–70.

Rupire J. (Last Checked 2015), La Cuicadora project, URL: http://cuidadora.ourproject.org/.

Free and Open Source Software for Geospatial (FOSS4G) Conference Proceedings

Volume 16 *Bonn, Germany* Article 7

2016

Is there life in Virtual Globes?

Konstantinos Evangelidis
Technological Educational Institute of Central Macedonia, Serres, Greece

Theofilos Papadopoulos
Technological Educational Institute of Central Macedonia, Serres, Greece

Recommended Citation

Evangelidis, Konstantinos and Papadopoulos, Theofilos (2016) "Is there life in Virtual Globes?," *Free and Open Source Software for Geospatial (FOSS4G) Conference Proceedings*: Vol. 16 , Article 7.
DOI: https://doi.org/10.7275/R56M3518
Available at: https://scholarworks.umass.edu/foss4g/vol16/iss1/7

Is there life in Virtual Globes?

Theofilos Papadopoulos, Konstantinos Evangelidis

Technological Educational Institute of Central Macedonia, Serres, GREECE – priestont@gmail.com, kevan70@gmail.com

KEY WORDS: Virtual Globes, 3D Geovisualizations, WebGL, Javascript

ABSTRACT:
Virtual globes have been widely used during last decade mainly for simulating observations of the earth from the outer space and navigation experiences over its surface which may be portrayed with various types of views and textures. The present work aims to extend virtual globe capabilities by incorporating three dimensional events on them. Such events may include animation and motion effects on 3D models representing real world living or inanimate spatial objects, modeling of natural resources and phenomena representations and any type of geovisualized activities and demonstrations.

1 Introduction

Virtual globes are browser-based web interfaces mainly used to offer to the user, navigation experience around the planet earth as being an observer at the outer space, with the ability in most cases to modify the observatory position and viewing angle. Various types of views are usually portrayed on the surface of the earth like for example ground terrain view, or road network map view or just aerial or satellite images of the real world. In addition, it is possible to represent three dimensional (3D) objects and convey on them types of textures. However, what is not being typically provided is animation and motion effects on 3D spatial entities for simulation and demonstration purposes.

The general idea behind the presented work is to extend current Virtual Globes capabilities by incorporating in their views, also motion and animation effects on 3D models representing living or inanimate spatial objects. Then to make it possible to create virtual geospatial worlds viewable on specified zoom levels on an area of the globe simulating various types of events such as historical, or related to flora and fauna, natural resources, phenomena etc.

What is needed to achieve the objectives described is a tool for developing layered-based geospatial scenes containing 3D actors participating in motions over a georeferenced area. Following that, a set of arrangements and specifications that will assure a proper appearance of all the above on a virtual globe. In this work, we make an attempt to develop a representative case of a living virtual globe displaying 3D scenes and events on the earth's surface and identify the development issues for such a venture.

2 Virtual Globes: Features and Trends

In the 00s, Riedl (2007) focused on the following Virtual Globe parameters: a) transportability, which refers to the capability of transporting the related geospatial data over the Internet, same way as it happens with any other digital information, b) scalability, as the capability of presenting geospatial information at any desired scale, c) interactivity, since Virtual Globes provide a web interface to receive feedback from end users, d) choice of topics as the capability of displaying unlimited number of global themes and e) currency, as the ability to present real-time or updated data at any given time span. Later on, Tuttle (2008), proposed two additional features: pseudo-3D, due to the naturally understood by people environment of Virtual Globes and client-side, which refers to the capability of end users to totally manage (insert, update, modify) the published data.

Since then, the geospatial technologies were boosted by the enormous progress occurred in information and communication technologies. We would choose as most significant, also affecting Virtual Globes, a) the developments in geospatial web semantics, which influenced data sharing, exchange and interoperability through XML based standards (KML, GML etc.) and OGC web services (WMS, CSW, WFS etc.) and b) the popularity of Javascript which triggered the development of specialized geospatial libraries.

A critical moment in Virtual Globes timeline was when they started to cooperate with Geospatial Web Services. According to Tuttle (2008) the simultaneous popularity of both technologies can not be a coincidence. In 2006 (Dunne and Shutton, 2006) discuss the powerful combination of Web Map Services with Virtual Globes. Today, it is possible to exploit geospatial services in the cloud in order to produce valuable geoinformation (Evangelidis et al, 2012) and Virtual Globes represent the mean for displaying this information.

As Javascript is nowadays the prevailing cross-platform programming language on the web compiled just-in-time on the client side, it is worth focusing on WebGL Javascript API, the standard for hardware accelerated interactive 3D graphics rendering on the web. Emerging visualization frameworks deployed on top of WebGL contribute significantly in the field of 3D geospatial developments, including those related to Virtual Globes, such as OpenWebGlobe7, WebGL Globe8 and Cesium.js9. An incomparable advantage of the above platforms is that they are directly executed by recent web browsers without the need of installing plug-ins or other software.

Cesium.js seems to have become the dominant geospatial visualization framework for 3D mapping on the web after Google announced its Earth API, deprecation (Angevine, 2015). It is built on top of WebGL, HTML5 standard, supports industry standard vector formats (KML, GeoJSON), Collada models when converted to its own glTF format, is open source and cross-platform, running in any browser supporting WebGL. It also supports 3D models animation and user-controlled motion over the terrain and numerous such examples were ported from Google Earth to Cesium (Fig. 1)

[7] http://www.openwebglobe.org/

[8] https://www.chromeexperiments.com/globe

[9] https://cesiumjs.org/

Figure 1: Motion and animation effects with Cesium.js.

However, the Javascript library much respected and recognized by the developers community which also exploits WebGL renderer capabilities, with numerous contributors and free examples for animation and extreme motion effects is that of three.js (http://threejs.org/). It supports complex geometries and highly detailed objects, however, to the best of our knowledge, it doesn't maintain the geospatial aspect of the involved scenes and virtual worlds. Three.js could act complementary to Cesium.js and vice-versa and their mixing might be the ideal case.

3 Development

On a previous work which is under review, we had developed on top of three.js a GIS based prototype web interface for creating custom geospatial worlds containing overlaid 3D objects moving on specified or dynamically user-defined motion paths and animated textures. To assist demonstration of the above mentioned work a video has been prepared and is available at https://goo.gl/mutZrs.

One of the obvious challenges and further developments was to integrate the above described "living" geospatial worlds with Virtual Globes. In the present work we made an attempt to perform such integration. The expected result is to view "life" over Virtual Globes after a reasonable zoom level.

The whole development process comprises the following discrete stages:

- The development of a tool for creating custom "living" geospatial worlds (already performed as previously demonstrated)
- The development of a function for creating Virtual Globes
- The development of a function for georeferencing custom geospatial worlds with the Virtual Globe

The whole development was exclusively based on three.js library.

3.1 Development of Custom Geospatial Worlds

A GIS digital elevation model of the demo area is created by providing a set of grid points with elevation values and by employing a plane type geometry[10]

10 http://threejs.org/docs/#Reference/Extras.Geometries/PlaneGeometry

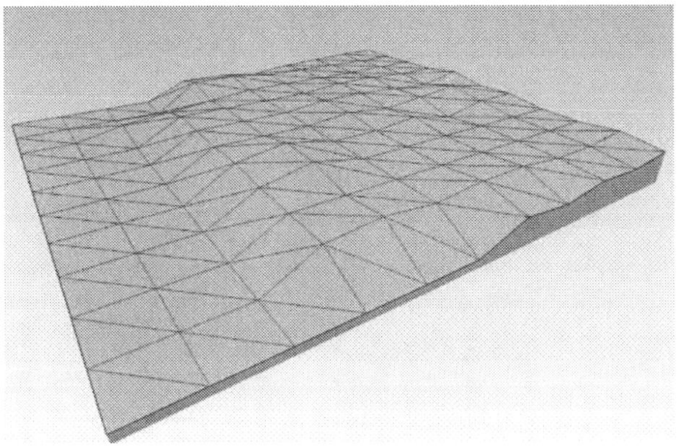

Figure 2: Digital elevation model created with plane type geometry of three.js

The texture of the geospatial world is loaded via a raster image which is adjusted appropriately

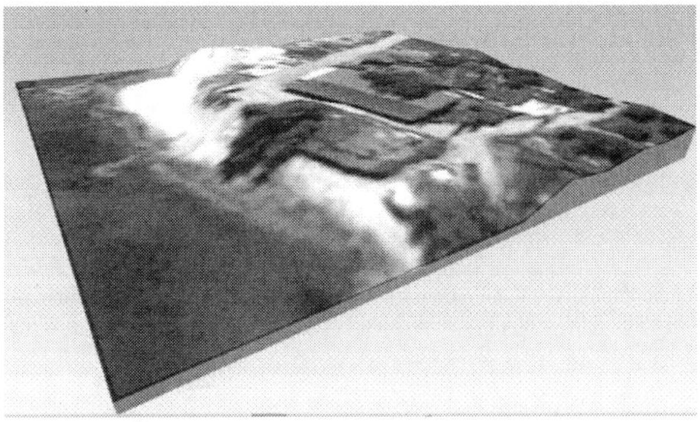

Figure 3: Surface texture adjusted over elevation model

Selected 3D models representing static spatial objects (e.g. house, plants) are scaled to fit in to the geospatial world. The static ones are placed in user-defined points, or along polylines or inside polygons as shown in figure 4.

Figure 4: Adding static 3D models placed on the geospatial world

Animated 3D models representing living (e.g. horse) or inanimate spatial objects (e.g. car) are also scaled and placed in user-defined motion paths specified by polylines. Animated textures of the surface (e.g. sea waves) are

also overlaid, as shown in figure 5

Figure 5: Overlaying motion 3D models on the geospatial world

3.2 Development of a Virtual Globe

To create a Virtual Globe in three.js using the WGS 84 main parameters the "THREE.ShpereGeometry" constructor is used with radius = 6378137 which refers to the major (equatorial) radius a in meters. To apply the polar semi-minor axis b, the flattening f=1/298.257223563 is used and applied to the scale method of the sphere geometry "earthGeometry.scale.set(1,1-f,1)" at the y axis (three.js axis are x,y – width,height and the z as the depth axis).

For the presentation purposes the Virtual Globe has been designed with its meridians and parallels every 10o degrees (Fig. 6) .

Figure 6: Virtual Globe created with three.js

3.3 Integrating geospatial world with Virtual Globe

In order to georeference the geospatial world with the Virtual Globe the following adjustments and acceptances should be considered:

- The custom geospatial world has to be designed or adjusted so that its axes are aligned with north-south and east-west orientations (y+/- : N/S, x+/-: E/W). In other words, the minimum x, minimum y values of the word should refer to the south-west of the custom world and the maximum x, maximum y to the north-east (Fig. 7).

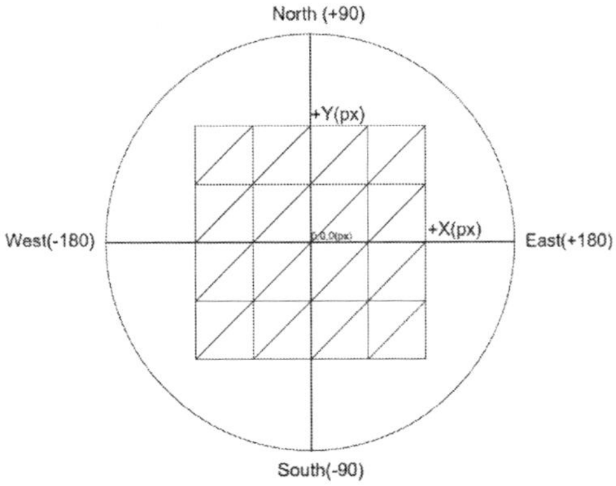

Figure 7: Adjusting custom world's axes orientation

- The center of the geospatial world is positioned at x:0, y:0) pixel coordinates. Once the geospatial world is georeferenced, then its center will be placed at its real WGS 84 φgw, λgw coordinate values.
- The center of the Virtual Globe is positioned at x:0, y:0 pixel coordinates and is placed at φo:0o, λo:0o WGS 84 coordinate values.
- Then, the Virtual Globe is rotated counterclockwise around x axis by φgw degrees and clockwise around y axis by λgw degrees. That way the x:0, y:0 pixel coordinates of the Virtual Globe are placed at the center of the geospatial world (Fig. 8 and Fig. 9).

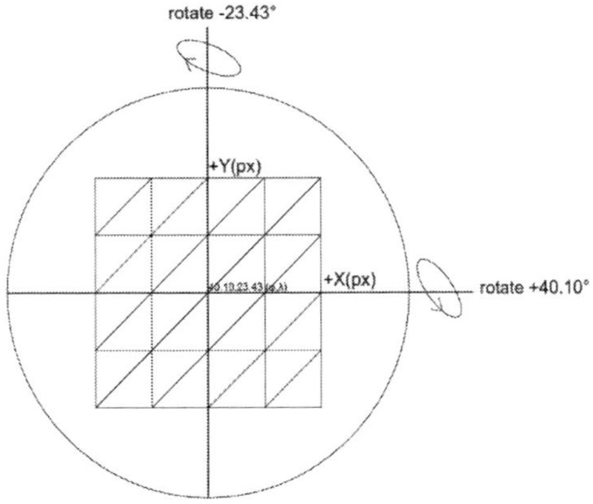

Figure 8: Rotating Virtual Globe to match geospatial world

As a result the 3D scene of the "living" geospatial world created is appropriately adjusted and is already viewable and may be explored through the Virtual Globe (Fig. 9)

Figure 9: Integrating "living" geospatial worlds with Virtual Globes

An indicative demonstration video presenting all of the above developments is available at http://goo.gl/M76uw3.

4 Conclusions & Further Developments

During last decade we experienced a significant progress occurred in the development of WebGL Javascript libraries, APIs and interfaces and their high penetration values on web applications. As regards the geospatial community, similar developments concentrate an ever increasing interest by geoscientists and professionals. Strong representative examples are found in Cesium.js for implementing Virtual Globes and 3D geospatial visualizations and model animations. Others are met in three.js developments for advanced 3D model animation and motion effects. To our intuition, the above Javascript libraries as well as new ones focusing also on geospatial data exchange and interoperability issues, are expected to dramatically affect Virtual Globes evolution and usability.

Challenges involve all Virtual Globes usability cases such as, the educational sector, through sophisticated large scale geo-visualizations containing motion and animation effects on areas sensible on geological phenomena and/or natural disasters or on regions of the atmosphere participating in meteorological processes.

So far, 3D rendering on the web for demanding imagery resolutions required special computer graphics hardware capabilities and high speed and quality internet connections. With the transition of processing on the client side and the adoption of open web standards and advanced interfaces (e.g. physics.js) it could become more efficient to perform complicated geovisualized simulations over the globe for typical engineering modeling: fluid mechanics, hydrological modeling for flood assessment, landslide assessments etc.

A virtual geospatial reality is just around the corner.

5 References

Riedl, A. (2007). Digital globes. In Multimedia cartography (pp. 255-266). Springer Berlin Heidelberg.

Tuttle, B.T., Anderson, S. and Huff, R. (2008). Virtual globes: an overview of their history, uses, and future challenges. Geography Compass, 2(5), pp.1478-1505.

Dunne, D., & Sutton, G. (2007). 3D Web-mapping: Integrating Marine Data into NASA World Wind. Hydro International, 7-9.

Evangelidis, K., Ntouros, K., & Makridis, S. (2012). Geoprocessing Services over the Web. In Proceedings of the 32nd EARSeL Symposium, Mykonos, Greece (pp. 344-349).

Angevine, G., Migrating from Earth to Cesium, [Online]. Available from: https://cesiumjs.org/2015/01/27/-Migrating-from-Earth-to-Cesium/

Free and Open Source Software for Geospatial (FOSS4G) Conference Proceedings

Volume 16 *Bonn, Germany* Article 9

2016

From global observations to local information: The Earth Observation Monitor

Jonas Eberle
Department for Earth Observation, Friedrich-Schiller-University Jena

Recommended Citation
Eberle, Jonas (2016) "From global observations to local information: The Earth Observation Monitor," *Free and Open Source Software for Geospatial (FOSS4G) Conference Proceedings*: Vol. 16 , Article 9.
DOI: https://doi.org/10.7275/R5Z31WVM
Available at: https://scholarworks.umass.edu/foss4g/vol16/iss1/9

From global observations to local information: The Earth Observation Monitor

Jonas Eberle [a,*]

[a] Department for Earth Observation, Friedrich-Schiller-University Jena – jonas.eberle@uni-jena.de

KEY WORDS: Web-based geo-processing; Earth Observation; time-series; vegetation analysis; web portal; mobile application; service-based infrastructure

ABSTRACT:

Earth Observation (EO) data are available around the globe and can be used for a range of applications. To support scientists and local stakeholders in the usage of information from space, barriers, especially in data processing, need to be reduced. To meet this need, the software framework "Earth Observation Monitor" provides access and analysis tools for global EO vegetation time-series data based on standard-compliant geo-processing services. Data are automatically downloaded from several data providers, processed, and time-series analysis tools for vegetation analyses extract further information. A web portal and a mobile application have been developed to show the usage of interoperable geospatial web services and to simplify the access and analysis of global EO time-series data. All steps from data download to analysis are automated and provided as operational geo-processing services. Open-source software has been used to develop the services and client applications.

1 Introduction

Information about environmental change is an important component in monitoring study areas on earth. Parameters of vegetation deliver valuable information about land cover and land use. Earth Observation satellites (e.g., Landsat, Terra, and Aqua) provide global datasets that are used for vegetation analysis. Time-series information of vegetation parameters delivers further information about changes, which can be used for new analyses. If the spatial resolution of these global datasets is suitable for local study areas, these datasets can also be used for local applications, giving local stakeholders the possibility of using global Earth Observation data for their specific needs. The transformation of global observations into local information facilitates the use of Earth Observation data for local studies, as well as the subsequent processing into local knowledge by involving stakeholders.

The involvement and consideration of stakeholders is an essential task in current international programs. A program launched in 2012, Future Earth (Future Earth 2013), was founded by international research funding organizations with the aim of supporting interdisciplinary collaborations in the area of global environmental change research, to address critical questions. In addition to questions regarding how the earth is changing, an important 'cross-cutting capability' is the focus on data and on observing systems, with the aims to "make the research more useful and accessible for decision makers" (Future Earth 2013:12) and "[make] research accessible to all parties" (Future Earth 2013:21).

Technical feasibilities are also mentioned in the Future Earth program seeking to determine best practices in integrating user needs and understanding research needs, by, for example "developing and diffusing useful tools for applying knowledge" (Future Earth 2013:21). On a more technical level, the Belmont Forum, a high level group of the world's major and emerging funders of global environmental change research and international science councils, aims to develop capable E-Infrastructures to increase community-level integration and collaboration, and to provide a community-owned global E-Infrastructure and Data Management Strategy and Implementation plan (Allison and Gurney 2014). The Group on Earth Observation (GEO) initiated the Global Earth Observation System of Systems (GEOSS; Group on Earth Observations 2005) with the aim of building a global public infrastructure that generates comprehensive, near-real-time environmental data, information, and analyses. To date, GEOSS has consisted of a metadata registry, a web portal, and a brokering system to tether external metadata catalogues to standard-compliant data services. However, only information can be downloaded from the metadata entries; it is often not possible to retrieve the data directly. Craglia et al. (2008)

published an agenda for the "Next-Generation Digital Earth" stating that multiple connected digital infrastructures and a problem-oriented focus to enable access to data, information, and services are needed, and visualization, open access, and multidisciplinary education and science is supported.

New insights into global environmental change, in focus with the next-generation digital earth, can be realized by making Earth Observation data and methods available with easy-to-use interfaces. An automated processing and analysis of Earth Observation time-series data is needed to fulfill this purpose. At present, many individual steps, such as discovery, download, pre-processing, analysis, post-processing, are needed to derive local information based on Earth Observation data. Only after completing these steps the user can derive the final information based on the outputs of the analysis tool, highlighting the need for ready-to-use derived information based on defined parameters. This can be achieved by operational and automated data processing to ensure easy access to this type of information.

The software framework "Earth Observation Monitor" (EOM) was initiated to provide easy access to Earth Observation time-series data and information from analysis tools, without the need to process data on the user's computer. The main users are local stakeholders, as well as scientists who need an overview of their area of study. The EOM makes applied analysis methods and tools developed by different research communities online available, in combination with data access tools and web services, all encapsulated within a straightforward web interface. The spatial data infrastructure developed deals with data search, download, and conversion, as well as analysis based on user inputs.

The objective of the present paper is to describe the functional and technical concept of the Earth Observation Monitor, as well as the applications developed. The EOM can be used to explore Earth Observation time-series data based on the following three principles:

- Simple time-series data access and analysis without data processing,
- Making software tools available via web services, and
- Providing simple tools with easy-to-use applications for stakeholders and scientists.

The remaining paper is structured as follows: Background and related work is shown in Section 2 with a focus on geospatial web services, vegetation time-series analysis, and available online tools. The design and implementation of the EOM are explained in Section 3 including data access and processing services as well as web and mobile applications. Finally, a summary and conclusions are given in Section 4 and future work in Section 5.

2 Background and related work

2.1 Geospatial service infrastructure

Modern web technologies facilitate the use of web services and are the basis for interactive and easy-to-use web-based geographic information systems. Using web services also allows the usage from multiple systems (e.g., a web portal, a mobile application, and other software clients). An important consideration in developing web-based geographic information systems is hiding the complexity of data processing techniques from users and providing simple web interfaces. Making desktop software available online relies on web services, which make it possible to execute tasks independently on remote servers. Client applications and service infrastructures can be separated (Figure 1) but interoperability needs to be integrated and provided; external clients can make use of the service infrastructure if access is permitted. Services are provided over the network, and thereby made available for client applications. Two main different architectural styles are available to publish services, service-oriented architecture (SOA; MacKenzie et al. 2006) and resource-oriented architecture (ROA; Overdick 2007). Both of them can also be used for geospatial services. Mazzetti, Nativi, and Caron (2009) have stated that "it is not possible to say that one architecture is better than the other" and that "the selection [...] depends on application requirements." These authors selected the REST approach for their Earth System Science applications.

In a spatial data infrastructure, services are available for the tasks of visualization, data access, metadata cataloguing, and data processing. To provide interoperable services and metadata, the Open Geospatial Consortium (OGC) published several standards and implementation specifications for the geospatial world. Those can be used to establish interoperability on server- and client-side. Automated and operational processing is needed for user-friendly time-series analysis tools, allowing users to perform the tasks of data downloading, processing, and analysis. The OGC has specified standardized processing services with the Web Processing Service (WPS; Schut 2007). This specification describes service interfaces that can be used to publish processing algorithms and to request their execution on the Web. Within the process algorithm, external software can be made available, for example, by executing software on the command line or through libraries used in programming languages. This allows a wide range of software to be integrated into a web service environment. Several open source software programs exist that meet OGC WPS specifications (an overview is given by Zhao, Foerster, and Yue 2012 or Lopez-Pellicer et al. 2012).

Web Processing Services are widely used in conjunction with spatial data infrastructures when processing methods are available to the user. Several publications also list any usage in research projects, and research has been done in the field regarding distributed processing (Friis-Christensen et al. 2007; Meng, Xie, and Bian 2010; Schaeffer et al. 2012; Foerster et al. 2011), semantic processing (Farnaghi and Mansourian 2013; Wosniok et al. 2014; Vitolo et al. 2015), process orchestration (Nash et al. 2007; Eberle and Strobl 2012; de Jesus et al. 2011; Wu et al. 2014; Meng, Bian, and Xie 2009), processing in the cloud (Sun 2013; Evangelidis et al. 2014), moving code (Müller, Bernard, and Kadner 2013), and modeling (Buytaert et al. 2012; Imran, Zurita-milla, and By 2011; Dubois et al. 2013).

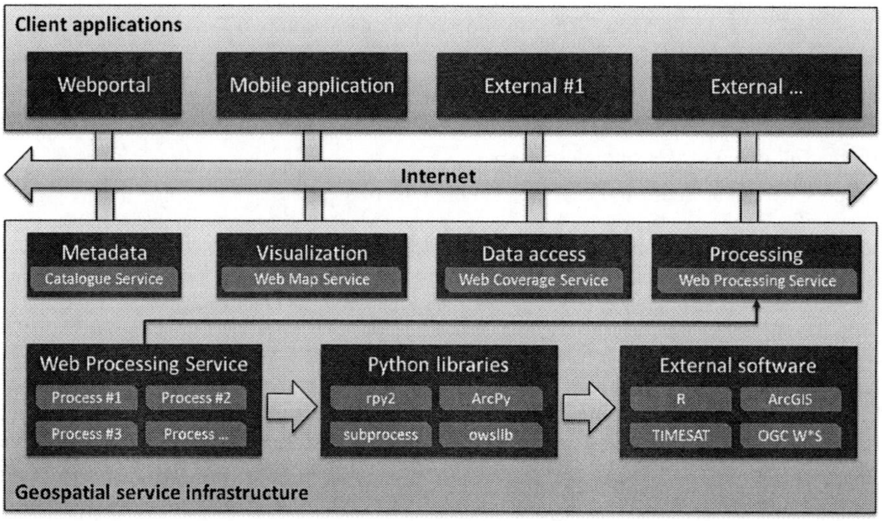

Figure 1: Client applications can connect through the internet to the geospatial service infrastructure that includes processing based on OGC Web Processing Services on Python programming language. Connections to diverse GIS processing software are made either through Python modules (e.g., R, ArcMap, OWS services) or through command line access (e.g., TIMESAT).

2.2 Vegetation time-series analysis

Satellite image time-series of vegetation indices are radiometric measurements of photosynthetically active radiation in the leaves of the vegetation, a proxy for the photosynthetic activity and vitality of plants. Vegetation indices like the Normalized Difference Vegetation Index (NDVI, Tucker 1979) or the Enhanced Vegetation Index (EVI, Huete et al. 2002) have proven to be important data sources for vegetation change and dynamics analyses. Daily to bi-weekly global data with a spatial resolution of 250m using the NDVI and EVI data from the Moderate Resolution Imaging Spectroradiometer (MODIS) allows vegetation analyses from 2000 onwards (Huete et al. 2002).

Evolving methods on vegetation change characterization have been developed by scientists around the world. Recent developments such as the 'Breaks For Additive Seasonal and Trend' (BFAST) method allow users to detect land cover changes by detecting phenological changes in the inter-annual time-series. BFAST integrates the decomposition of time-series into trend, seasonal, and remainder components, and provides the time and number of changes in the time-series. Studies observing increasing or declining net primary productivity of vegetation (Verbesselt, Hyndman, Newnham, et al. 2010; Verbesselt, Hyndman, Zeileis, et al. 2010) fostered the development of statistical tools for detecting trends such as greening or browning. The greenbrown package (Forkel et al. 2013) is a collection of functions designed to analyze trends and trend changes in gridded time-series, like those from satellite observations or climate model simulations. In the Earth Observation community, Greening describes positive trends in vegetation greenness, whereas Browning indicates negative trends in vegetation greenness. Both methods are provided as tools in the R statistical language. In addition to trend analyses of satellite time-series, the derivation of phenological metrics for vegetation characterization and classification is another important issue. Introduced by Jönsson and Eklundh (2004), the software TIMESAT has been developed for satellite-based phenological analyses based on the seasonality of satellite time-series data and their relationship with the phenological cycling of vegetation. Further on, it enables the extraction of time-related phenological metrics (e.g., start of season, length of season) or biomass-related metrics, such as the integral of the vegetation period.

2.3 Online time-series access and analysis tools

A number of tools for Earth Observation time-series data access and analyses are available online, but they have all specific limitations, described in this chapter. More and more tools allowing time-series extraction and analysis of specific datasets from NASA MODIS sensor are becoming available. Some of these tools are listed in Table 1, and are discussed in the following paragraphs.

The probably best known web-based tool is NASA Giovanni – Interactive Visualization & Analysis (Acker and Leptoukh 2007), which offers a wide range of available MODIS datasets and different time-series extraction and analysis tools. Although NASA Giovanni does provide access to all input data for a given bounding box and time range, these tools are not publicly available as web services to be integrated into external applications. The MODIS Land Subsets tool from the United states' Oak Ridge National Laboratory (ORNL) provides time-series extraction for a bounding box around a selected pixel location. In addition to time-series data access, the tool also visualizes land cover types and provides a phenology time-series plot for vegetation data. A SOA-based MODIS Web Service is publicly available for querying the data catalogue and submitting an order for time-series access and analysis graphs. The Laboratory of Remote Sensing in Agriculture and Forestry of the Brazilian National Institute for Space Research (INPE) developed a tool for MODIS EVI-2 and TRMM precipitation data plotting for South America; an easy-to-use web portal provides time-series data with just one click on the map. The Multi-Sensor Evolution Analysis (MEA) of the EU EarthServer project focuses on the provision of time-series data access, based on OGC-compliant web services. Through the web portal, tools are available for interactive time-series plotting for different pixels and datasets, which can be compared using a chart. Spatial subset plots can also be shown as images or as map-based visualizations. Via the OGC Web Coverage Processing Service (WCPS), the datasets are available in a web service query-processing language.

Table 1: Comparison of web-based time-series data access and analysis tools.

	Outputs / Web Services
NASA Giovanni http://giovanni.gsfc.nasa.gov/ Available portals: Atmospheric, meteorological portals Ocean & hydrology portals	Results as images, GeoTiff files, ASCII text. GeoTiff files can be visualized interactively with an OGC Web Map Service.
MODIS Land Subsets http://daac.ornl.gov/MODIS/ Available MODIS products (excerpt): BRDF, Surface Reflectance Land Surface Temperature Vegetation Indices, Leaf Area Index	Results as interactive and image plots, GeoTiff files, ASCII text. MODIS Web Service based on SOAP for data queries and access (only ordering).
INPE EVI2 tool https://www.dsr.inpe.br/laf/series/en/ Available data: MODIS EVI-2, SRTM, TRMM	Results as interactive plots, CSV export.
Earthserver MEA http://earthserver.services.meeo.it/ Available MODIS products: Atmospheric datasets MOD13Q1 NDVI	Results as interactive plots, XML export. Datasets available as OGC Web Coverage Service (WCS) and Web Coverage Processing Service.

For users focusing on vegetation time-series data, many of the tools described in the preceding paragraphs offer web-based tools for time-series extraction and visualization. However, none of the existing tools can yet be integrated into external applications. The OGC WCPS service from MEA is a good approach to overcome this issue; however, the complete MODIS vegetation datasets with Enhanced Vegetation Index, the Normalized Difference Vegetation Index, and the quality flag datasets are not yet fully available and are not yet updated automatically. Although the MODIS Web Service from ORNL provides users with a way to submit order requests, the resulting data are only available on a website whose link is sent via electronic mail. To sum, service-based access to time-series vegetation data is only available within specific limits. Using OGC Web Processing Services is therefore a good technical solution, allowing users to execute the services directly within other external applications.

From the point of view of an end-user, further information based on the analysis of time-series data is also useful. Some of the scientific algorithms that are currently available can provide information about breakpoints in trends and seasonality that can be used, for example, as indicators for changes in vegetation. A great deal of information can also be retrieved from phenological derivations, like the start and end of a season, seasonal amplitude, and seasonal integral. However, to date, none of this information is available as on-demand processing services in web-based environments.

3 The Earth Observation Monitor

The Earth Observation Monitor (EOM) is a software framework with the aim of ensuring easy access and analysis of spatial time-series data for land monitoring on local scales. Scientists and stakeholders working in local areas need to be able to do time-series analysis without the need to process any data. In this chapter, the concept; technical developments based on automated data access, processing, and analysis; and example client applications are described.

3.1 Concept

The concept behind the EOM combines the advantages of web service-based geoprocessing with user-friendly interfaces; this provides an easy access for users without specific knowledge in data processing. The EOM focuses on hiding existing barriers (e.g., data download, data processing, format conversion), to allow users to focus on the analysis results. As many steps that are needed to produce analysis results from Earth Observation data can be automatized, users can concentrate on the interpretation of the analysis results. Figure 2 illustrates the tasks: The EOM integrates the data for the specified area including quality checks and data clipping. If the analysis method and its parameters are set, the system prepares data needed for the analysis, executes the method, and prepares the results for online visualization and easy-to-use offline usage.

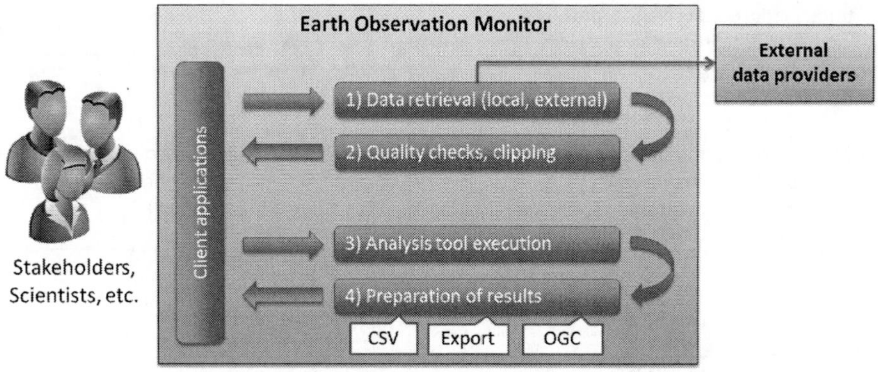

Figure 2: Concept of the Earth Observation Monitor.

3.2 Automated data access

The basis for an automated data processing and analysis tool is operational and automated data access, which is accomplished by introducing a multi-source data processing middleware (Eberle et al. 2013). This middleware (see Figure 3) integrates data from external sources (like NASA, NOAA, Google Earth Engine, and general datasets) on demand. Any integrated dataset is described with metadata and stored as geospatial data in a common data format. These data are provided to client applications through OGC-compliant web services, the integration is also available as OGC Web Processing Service.

The middleware was developed as a Python library named pyEOM . The aims of the library are (1) to provide access to time-series data as easily as possible and (2) to provide a standard-compliant web service for data access, visualization, processing, and analysis. This library can either be used within a Python script or by the web processing services published by the EOM framework. pyEOM provides access to several external datasets like MODIS and climate station data. Table 2 lists available data sources. Datasets from OGC Web Coverage Services (WCS) and OGC Sensor Observation Services (SOS) can also be integrated into the middleware. Any dataset integrated into the middleware can be visualized with OGC-compliant web services. Point-based time-series data can be published as an OGC SOS; Polygon-based time-series raster data is available as an OGC Web Map Service (WMS) for visualization and as OGC WCS for data download.

Table 2: Datasets connected in the multi-source data processing middleware (adjusted after Eberle et al. 2013).

Dataset	*Type*	*Spatial res.*	*Temporal*	*Time range*
NASA Land Processes Distributed Active Archive Center (LPDAAC)				
MODIS	Raster	250m – 1km	Daily to Monthly	2000 – now
National Snow and Ice Data Center (NSIDC)				
MODIS	Raster	250m – 1km	Daily to Monthly	2000 – now
NOAA National Climatic Data Center				
GSOD, GHCN	Sensor	-	Daily	1929 – now
Google Earth Engine				
MODIS	Raster	250m – 1km	Daily to Monthly	2000 – now

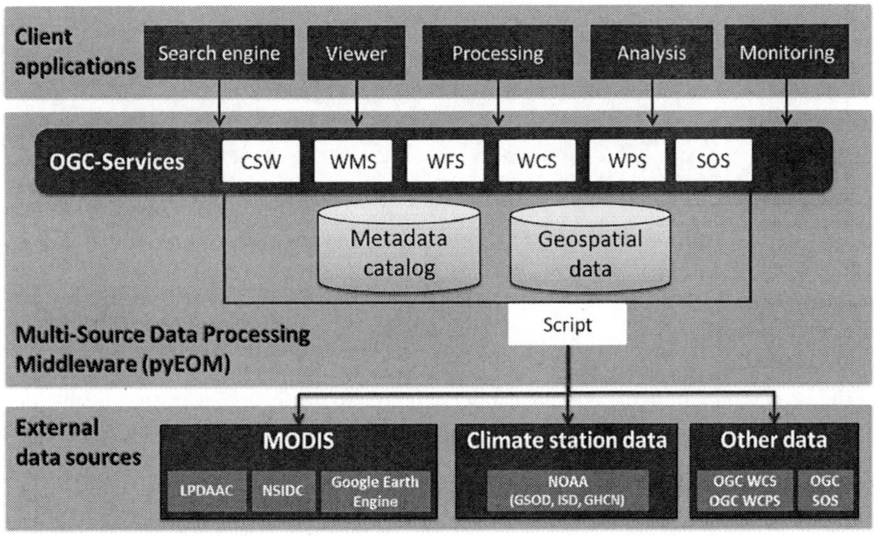

Figure 3: The Multi-Source Data Processing Middleware (updated after Eberle et al. 2013) is connected to external data sources and provides data integrated into the middleware as OGC-compliant web services. Geospatial data is stored in a common data format, meta

3.3 Automated data processing and analysis

Data processing and analysis need to be available via a web-based system. To make processing and analysis tools automatically available, they must be executable either within a programming language or from the command line. *PyWPS*[3] was used as Python-based open-source software compliant with the OGC Web Processing Service specifications. Ready-to-use Python-libraries, such as *rpy2* for R statistical language or *ArcPy* for ArcMap tools, as well as command line executions (with the Python library subprocess), can be used to run external software for data analysis or data processing (see Figure 1).

The automated workflow can be divided into five steps:

1. Time-series data extraction for a point or polygon selected by the user.
2. Data post-processing for quality checks and interpolation.
3. Data preparation for the individual analysis tool.
4. Execution of the analysis tool.
5. Preparation of analysis results for online visualization and download.

For any processed dataset of a created feature-a point or a polygon-,an individual processing directory is created that can be accessed by the user. Within this processing directory, any downloaded and processed data is available for download. Any input data, as well as results for a selected analysis tool, are also stored in this directory (see Figure 4). For example, the user can download the input data to execute the analysis tool offline on the local computer to reproduce the results. The individual steps of the automated geoprocessing services (Figure 4) are described in the sections 3.3.1 and 3.3.2. Appendix B lists all WPS processes that have been developed for data retrieval and data analysis.

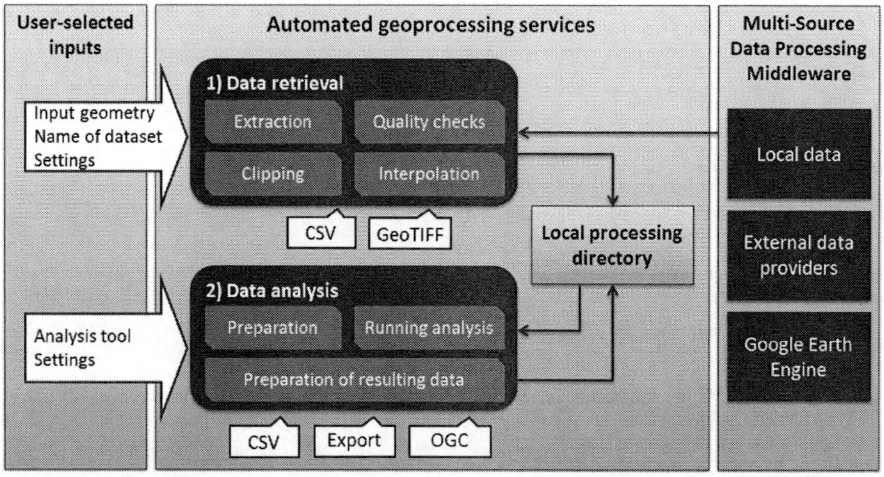

Figure 4: Automated data processing and analysis workflow.

3.3.1 Data retrieval

The retrieval of data is the basis for all further analyses. Dependent on the selected dataset, feature geometry, and source of the dataset, different processing steps have to be conducted. The processing includes the extraction of time-series data from the original dataset, clipping to the feature geometry, quality checks based on user inputs, and linear interpolation to replace `no-data` values from the time-series. To ensure transparency, quality flags and information regarding any interpolated values are stored. The data retrieval has been made available as OGC WPS process to be able to retrieve data within external client applications.

Figure 5 shows an example request for the WPS process to integrate a MODIS MOD13Q1 EVI dataset or a lat/lon coordinate. Further parameters can be set to consider a specific time range and quality flags. The request can be executed synchronously or asynchronously (including the parameters `status` and `storeExecuteResponse`). The resulting output consists of files including data and plots as well as a unique identifier (uuid). This identifier is needed afterwards in the execution of analyses processes to reference the data integrated in this step.

[3] http://pywps.org/

```
http://artemis.geogr.uni-jena.de/cgi-bin/testbox.cgi?
    service=WPS&version=1.0.0&request=Execute&
    identifier=1013_single_ts_plot_point&
    datainputs=[
        datasetName=mod13q1_evi;
        pointX=13.54;
        pointY=52.31
    ]
    &status=true&storeExecuteResponse=true          (optional)
```

Figure 5: Request to integrate a MOD13Q1 EVI dataset for a given coordinate.

Extraction & Clipping

In the first step of data retrieval, the required information needs to be extracted out of the raw time-series dataset and clipped to the user's geometry. Dependent on the geometry type (point or polygon), the output of the common data format differs: For points, a CSV file with the time-series information is provided; for polygons, GeoTiff files are stored individually for each timestamp.

Quality checks

Products from MODIS sensor are always delivered with quality flags so that users can decide which data are good enough for their specific application. To give users the opportunity to determine data quality, quality flags have to be processed. Depending on the threshold being used, some data can be declared as having no-data value. The quality flags are also extracted and clipped to the users' geometry. A quality mask is then created for each timestamp, based on the users' quality parameter inputs. These masks are then applied to the corresponding raster data.

Interpolation

For some time-series analysis methods missing data need to be interpolated, if no-data values are present. A linear interpolation is used to overcome this issue in the pre-processing step. Interpolated values are marked in a separate column in the final output CSV file, as well as in the output plot.

3.3.2 Data analysis

The basis for the time-series data analysis is either a CSV file, for point-based analyses, or a multi-band raster image stack, for polygon-based analyses. The following steps concerning data preparation, analysis execution, and preparation of resulting data are mostly individual for each analysis tool; these steps are also different for point-based and polygon-based analyses. As with data retrieval, any processed data is stored in the processing directory that is available to the users, giving them the possibility to download the prepared data for a specific analysis tool and to reproduce the analysis offline on their computers. The OGC WPS processes developed for time-series analysis are listed in Appendix B. Two processes always exist for each analysis method, one for point-based and one for polygon-based analyses.

Figure 6 shows an example request for the WPS process to execute BFAST for a prior integrated dataset. The dataset is specified by the unique identifier (uuid) as output from the data retrieval process. Further optional parameters can be set to adjust the BFAST algorithm. The resulting output consists of several plots and files (see Table 3).

```
http://artemis.geogr.uni-jena.de/cgi-bin/testbox.cgi?
    service=WPS&version=1.0.0&request=Execute&
    identifier=2010_single_ts_bfast_point&
    datainputs=[
        uuid=b5b17389-05d7-4058-9d87-aea05a0234d7
    ]
    &status=true&storeExecuteResponse=true          (optional)
```

Figure 6: HTTP request to execute bfast for a prior integrated dataset (defined by the uuid parameter).

Data preparation

The preparation of datasets for specific analysis tools can be automated based on the requirements of the selected tool. For the R-packages 'BFAST' and 'greenbrown' it is necessary to build a time-series object with a starting date and a frequency parameter to describe the time-series. For polygon-based analyses, both packages need a *raster brick* object that can be created based on the multi-band raster file. Data pre-processing for TIMESAT consists of the creation of a configuration file and the preparation of the input data. For point-based analyses, the input data is a new ASCII text file with data values in a single row. For polygon-based analyses, the data must first be scaled to the original values and converted into the ENVI HDR raster data format, which does not include header information. This is done with the Geospatial Data Abstraction Library (GDAL; (Warmerdam 2008). Afterwards, a list of files is saved in an ASCII file that is referenced in the TIMESAT configuration file.

Analysis execution

The execution of analysis methods can be done either from the command line or using a Python library. The Python library rpy2[4] was used as a direct connection to R within Python. With this library it is possible to execute R functions and to exchange variables between both languages. The execution of TIMESAT is done from the command line, which is executed from the subprocess module of Python. In such an automated environment, no command line prompts can occur; rather, the complete execution has to run in a single step. The resulting data need to be stored in files that can be accessed by the Python process after successful execution.

Preparation of analysis results data

The resulting data can then be prepared to ensure easy user access. A graphical figure can be made from the results of any analysis method; several output files are also generated, depending on the feature geometry and the analysis method. Table 3 lists the prepared output files and services. To visualize spatial analyses of feature polygons, OGC Web Map Services (WMS) and OGC Web Feature Services (WFS) are automatically generated from the raster and vector outputs: GeoTiff files are provided as WMS, and shape files as WFS. The shape files show each pixel of the polygon in the map, and come with an attribute table, which gives the statistics of each pixel by analysis method. This attribute table can also be obtained using the OGC WFS GetFeatureInfo request.

Table 3: Generated output files, by feature geometry and analysis method.

	BFAST	**greenbrown**	**TIMESAT**
Point	Data (CSV) Statistics (CSV) Plot (PNG)	Data (CSV) Statistics (CSV) Plot (PNG)	Data (CSV) Phenology (CSV) Plot (PNG)
Polygon	Data (GeoTiff) OGC WMS Pixel-Shapefile	Classification, slope, p-Value (GeoTiff, WMS) Shapefile / WFS for slope, p-Value	Data (GeoTiff) OGC WMS Shapefile / OGC WFS for Phenology

3.4 Client applications

The functions of the EOM are available with web services, allowing client applications to make use of these services. Two example clients were developed to show the possibilities of such a service-based infrastructure: A web portal (webEOM) and a mobile application (mobileEOM). Both of these clients use the OGC WPS for data integration and analysis. OGC Web Map and Feature Services for resulting data provide the possibility for direct online visualization.

3.4.1 webEOM

The focus of the web portal is to provide an easy-to-use client, while making it possible to extract time-series data and execute further time-series analysis functions. The webEOM[5] map (Figure 7) can be used to create the geometry of a study area. Based on this geometry, the system requests a list of available datasets that are registered in the multi-source data processing middleware. When conducting data integration, users can specify different parameters for the selected dataset, such as start and end dates, as well as filtering options. A processing directory for each integrated dataset is available to users, which contains any processed data for both the dataset and the executed analysis tools. A time-series and a decomposition plot are generated automatically

[4] http://rpy.sourceforge.net

[5] http://www.earth-observation-monitor.net/map.php

from the extracted time-series data that are plotted in the web portal. In addition to the images, CSV files are created and available for download. After data integration, users can select an analysis tool for execution, and individual parameters can be set for the selected analysis. The resulting data can either be visualized directly in the web portal or can be downloaded for further usage. Spatial outputs can be interactively explored in the map, and CSV files will be plotted as an interactive chart.

Several open-source libraries were used to develop the web portal (Appendix A): Drupal CMS is used for user registration and authentication, as well as as a registration service for created features. On the frontend interface, the JavaScript library OpenLayers is used as a mapping library; dygraphs are used to generate interactive charts; and jQuery provides standard JavaScript functions.

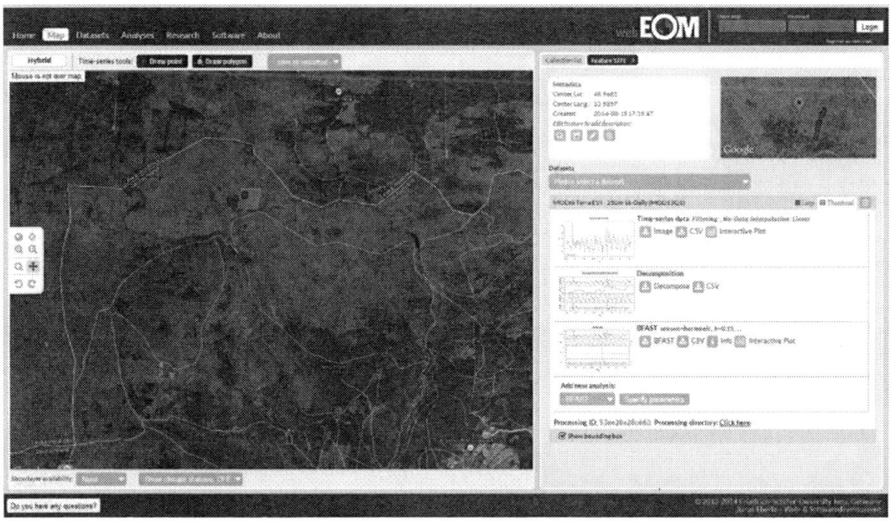

Figure 7: Screenshot of webEOM portal.

3.4.2 mobileEOM

During fieldwork, users cannot use web-based systems developed for desktop computers. A mobile application is therefore needed to foster the use of spatial time-series tools on mobile devices, which can be more easily used during work in the field. With such a mobile application users have access to a wide range of information, because a time-series of several years (15 years for MODIS vegetation data) can tell stories about the changing environment. The combined usage with time-series analysis tools allows identifying breaks in the past and overall trends of the vegetation, while users are in the field.

The mobile application for EOM was developed to provide access to time-series data and derived analyses on mobile devices. Using their current GPS location or a manual set position, users can extract vegetation time-series data, as well as view data plots, trend, and breakpoint analysis plots directly on their mobile devices. An OGC WPS process was developed for the mobile application. This provides all necessary functionalities in a single process, available as web service. This process extracts the data from Google Earth Engine[6] and plots the time-series and decomposition figure. In a second step, time-series analyses for breakpoint detection (BFAST) and trend calculations (greenbrown) are executed and plotted in a figure. The resulting output is a GeoJSON file containing the values of the analysis tools, as well as links to the generated figures. Figure 8 shows screenshots of the mobile app linked with a chart of the OGC WPS process and how they interact.

[6] https://earthengine.google.org/

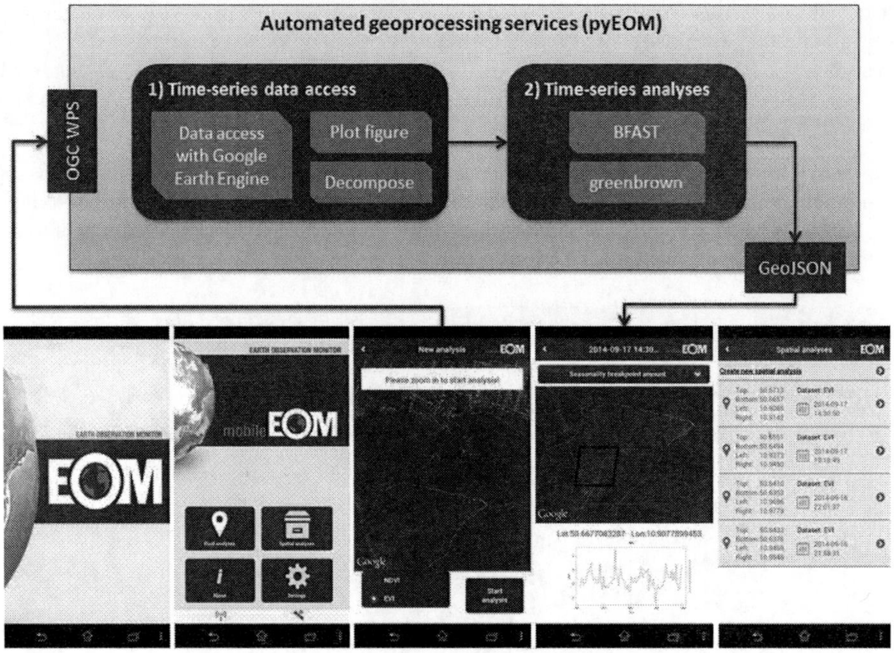

Figure 8:Screenshots of the mobileEOM app and interactions between the app and EOM web services.

The mobile application was developed as hybrid web application. Applications of this type are developed using HTML5, CSS, and JavaScript and then exported as native applications. To access the sensors and functions of the mobile device (e.g., GPS, file storage), the software PhoneGap was integrated and used for native application export. The interface was developed using jQuery Mobile, and the interactive map was integrated with Google Maps API.

4 Summary and conclusions

Based on an infrastructure using standard-compliant geospatial web services, the Earth Observation Monitor (EOM) software framework was introduced to provide global access and analysis tools for NASA MODIS vegetation time-series data, for user-defined points and polygons. A further aim was to make such time-series data and algorithms developed in research projects for data analysis more useful and accessible for users (i.e., scientists, stakeholders, and other decision-makers). Interoperable web-based geoprocessing services have been developed that can be used from external software and linked from data registries and metadata catalogues (e.g., GEOSS portal). Metadata entries can now be linked with such data-access services as provided within the framework of EOM. Two client applications have been developed to show how EOM can provide simple access to these services via a web portal and a mobile application, meaning that users do not need to process any data themselves.

Earth Observation time-series data are now operationally available for a wide range of applications. In many cases, users are interested in information about the change of the environment in their specific area of interest. Valuable information can be derived from time-series data for such user-defined areas, but better tools are needed to extract time-series data and to provide this information. With increasingly large data archives and high volumes of requests for data access, standard-compliant services for data access must be provided. In addition to traditional ways of providing data access services from the Open Geospatial Consortium (OGC), like Web Feature Service and Web Coverage Service, the present article proposes using Web Processing Services (WPS). Individual data integration and processing steps can be realized using WPS, such as applying individual quality masks, sub-setting data to a specific polygon, and providing different output formats. Until today only a few WPS processes are available for data access and processing; these need to be considered also by data providers.

Diverse scientific subjects (e.g., biodiversity monitoring, conservation ecology, and forest management) can benefit from using global Earth Observation time-series data, but data access still remains a complex task (Kuenzer et al. 2014). In addition to needing easier data access, interoperable analyses methods must be integrated to enable further information derivation. Global data can be transformed into locally relevant information by extracting datasets for local areas and applying algorithms. For time-series data, these algorithms can be land-surface trend calculations, breakpoint detection, or phenological analyses of vegetation data. Providing access to such algorithms—for example, by implementing web services for algorithms that have been developed within research communities—will make this area of research more accessible. This was shown in the present article for vegetation time-series analysis, and led to the Earth Observation Monitor framework.

An ideal way to make data access easier is providing web services, as they can be easily integrated into other applications and used within programming languages. The OGC WPS specification provides a standardized framework for geospatial processing, allowing other external applications to use these services with a common interface. Only at least the name of a dataset and the geometry (point, polygon) are necessary to access the data; the user does not need to know the details in handling specific Earth Observation datasets. As Earth Observation data are mostly provided in diverse data formats, these web services can provide consistent data formats, which are easier for users. In addition to providing easier data access, information based on time-series data must also be made more available. By providing scientific algorithms as web services, both algorithms and time-series data can be linked. This linking then allows users to execute both the integration process for data access and the algorithm process for information extraction. Without needing to do any data processing at all, users can retrieve locally relevant information, such as climate-driven vegetation phenology trends and back-dating of land-cover change events using breakpoints.

Web clients and mobile apps can be used in conjunction with geospatial web services, increasing the effectiveness and interoperability of web services by adding an intuitive frontend interface for different devices. Such an interface can, for example, provide notifications when data are available for analysis. A mobile app gives users direct access to time-series data while they are conducting fieldwork, without requiring a notebook or desktop computer to have access to the time-series data. Such applications can be built on top of the developed web services, and, as they are available over the Internet, can be used from any application that is compliant with the OGC WPS specifications.

5 Future work

Although the technical solutions have shown that geoprocessing services can be used to lower barriers in time-series data acess and analysis, there are still some topics that can be considered in the future: The processing services need to be described with metadata and made available in catalogue services and system of systems (e.g., GEOSS). Inputs and outputs of a process need to be semantically described to automatically detect what kind of inputs can be used with a process and to establish automatic connections between processes as workflows. Also the "moving code" (Müller et al. 2013) paradigm needs to be considered further to make it easier to process algorithms on external servers.

The further development of the EOM is concentrates on the usage of distributed processing capabilities to increase the performance and the integration of further high- and medium-resolution satellite sensors, such as the new Copernicus Sentinel satellites or Landsat. Furthermore the web and mobile applications can be optimized in various ways.

Acknowledgements

The development of the Earth Observation Monitor has been funded by Friedrich Schiller University, Jena, and by EU FP7 EuRuCAS project (No. 295068). The authors thank NASA and USGS for providing free and open access to their MODIS data archives, as well as the Google Research Team for providing support and access to Google Earth Engine. Additional thanks go to Franziska Zander (Friedrich Schiller University, Jena, Germany) for proofreading the paper.

Appendices

A. List of software used for EOM services, webEOM, and mobileEOM

Software	Version	Function

Drupal CMS	7	Backend
jQuery	1.10.2	JavaScript Frontend Library
OpenLayers	2.12	JavaScript Mapping Library
dygraphs	1.0.1	JavaScript Chart Library
Google Maps API	3	JavaScript Mapping Library
jQuery Mobile	1.3.2	JavaScript Mobile Library
PhoneGap / Cordova	3	Native Mobile Application Framework
MapServer	6	OGC WMS & WFS output layers for time-series analysis results
PyWPS	3.2.2	OGC Web Processing Service
R	3	Time-series analysis (BFAST, greenbrown)
Python	2.7	Libraries such as GDAL, OGR, Pandas
TIMESAT	3.1	Phenological time-series analysis

B. Web processing services for time-series data retrieval and time-series analysis available for EOM: http://artemis.geogr.uni-jena.de/cgi-bin/testbox.cgi

Data access for points	
Identifier:	1012_single_ts_plot_point
Inputs:	pointX, pointY, datesetName, dateStart, dateEnd, qualityLimits
Outputs:	pointBBOX, tsCSV, tsPlot, tsDecompose, tsDecomposeCSV, uuid
Data access for polygons	
Identifier:	1012_single_ts_plot_polygon
Inputs:	wkt, datesetName, dateStart, dateEnd, qualityLimits
Outputs:	tsCSV, tsPlot, tsDecompose, tsDecomposeCSV, uuid
Data access for climate stations	
Identifier:	1020_single_ts_plot_climate
Inputs:	station, parameters, dateStart, dateEnd
Outputs:	tsCSV, tsPlot, uuid
Time-series BFAST analysis	
Identifier:	2010_single_ts_bfast_point | 2010_single_ts_bfast_polygon
Inputs:	uuid, season, h, maxiter, breaks
Outputs:	Point: seasonalityDates, trendDates, bfastCSV, bfastPlot Polygon: layers, plot
Time-series greenbrown analysis	
Identifier:	3010_single_ts_greenbrown_point | _polygon
Inputs:	uuid, method, mosumpval, h, funSeasonalCycle, funAnnual
Outputs:	Point: breakpoint, slopes pvalue, gbCSV, gbPlot, trendCSV Polygon: layers, mac, mac_cl, mac_plot
Time-series TIMESAT analysis	
Identifier:	4010_single_ts_timesat_point | _polygon

Inputs:	uuid, fitting, seasonStartMethod, seasonStart, seasonEnd, spikeMethod, spikeValue, adaptionStrenght, windowSize, noIterationsUpperEnvelope
Outputs:	Point: timesatFitCSV, timesatSeasonalityCSV, timesatPlot Polygon: layers, seasonalityParameters, seasons
Data availability	
Identifier:	0020_wps_available_datasets
Inputs:	wkt (Point or Polygon representation as string)
Outputs:	json (list of datasets)
Mobile point-based WPS	
Identifier:	8010_single_ts_mobile
Inputs:	pointX, pointY, qualityLimits, uuid, width
Outputs:	pointBBOX, tsCSV, tsPlot, tsDecompose, tsBfast, tsGreenbrown, uuid
Mobile point-based WPS	
Identifier:	8020_multiple_ts_mobile
Inputs:	bbox_top, bbox_right, bbox_bottom, bbox_left, dataset, width
Outputs:	results, analysis, uuid

6 References

Acker, James G., and Gregory Leptoukh. 2007. "Online Analysis Enhances Use of NASA Earth Science Data." *Eos, Transactions American Geophysical Union* 88 (2): 14.

Allison, Lee, and Robert Gurney. 2014. *Belmont Forum E- Infrastructures and Data Management Collaborative Research Action.*

Buytaert, Wouter, Selene Baez, Macarena Bustamante, and Art Dewulf. 2012. "Web-Based Environmental Simulation: Bridging the Gap between Scientific Modeling and Decision-Making." *Environmental Science & Technology* 46 (4): 1971–76.

Craglia, Max, Michael F Goodchild, Alessandro Annoni, Gilberto Camara, Michael Gould, Werner Kuhn, David Mark, Ian Masser, and David Maguire. 2008. "Next-Generation Digital Earth." *International Journal of Spatial Data Infrastructures Research* 3: 146–67.

De Jesus, J., P. Walker, M. Grant, and S. Groom. 2011. "WPS Orchestration Using the Taverna Workbench: The eScience Approach." *Computers & Geosciences*, November. Elsevier, 1–12.

Dubois, G., M. Schulz, J. Skøien, L. Bastin, and S. Peedell. 2013. "eHabitat, a Multi-Purpose Web Processing Service for Ecological Modeling." *Environmental Modelling & Software* 41 (March). Elsevier Ltd: 123–33.

Eberle, Jonas, Siegfried Clausnitzer, Christian Hüttich, and Christiane Schmullius. 2013. "Multi-Source Data Processing Middleware for Land Monitoring within a Web-Based Spatial Data Infrastructure for Siberia." *ISPRS International Journal of Geo-Information* 2 (3): 553–76.

Eberle, Jonas, and Christian Strobl. 2012. "Web-Based Geoprocessing and Workflow Creation for Generating and Providing Remote Sensing Products." *Geomatica* 66 (1): 13–26.

Evangelidis, Konstantinos, Konstantinos Ntouros, Stathis Makridis, and Constantine Papatheodorou. 2014. "Geospatial Services in the Cloud." *Computers & Geosciences* 63 (February). Elsevier: 116–22. doi:10.1016/j.cageo.2013.10.007.

Farnaghi, Mahdi, and Ali Mansourian. 2013. "Disaster Planning Using Automated Composition of Semantic OGC Web Services: A Case Study in Sheltering." *Computers, Environment and Urban Systems* 41 (September). Elsevier Ltd: 204–18.

Fielding, Roy Thomas. 2000. "Architectural Styles and the Design of Network-Based Software Architectures." University of California, Irvine.

Foerster, Theodor, Bastian Schäffer, Bastian Baranski, and Johannes Brauner. 2011. "Geospatial Web Services for Distributed Processing - Applications and Scenarios." In *Geospatial Web Services: Advances in Information Interoperability*, edited by Peisheng Zhao and Liping Di, 245–86. Hershey, PA, USA1.

Forkel, Matthias, Nuno Carvalhais, Jan Verbesselt, Miguel Mahecha, Christopher Neigh, and Markus Reichstein. 2013. "Trend Change Detection in NDVI Time Series: Effects of Inter-Annual Variability and Methodology." *Remote Sensing* 5 (5): 2113–44.

Friis-Christensen, Anders, Nicole Ostländer, Michael Lutz, and Lars Bernard. 2007. "Designing Service Architectures for Distributed Geoprocessing: Challenges and Future Directions." *Transactions in GIS* 11 (6): 799–818.

Future Earth. 2013. *Future Earth Initial Design: Report of the Transition Team*. Paris, France. http://www.futureearth.info/sites/default/files/Future-Earth-Design-Report_web.pdf.

Group on Earth Observations. 2005. "The Global Earth Observation System of Systems (GEOSS) 10-Year Implementation Plan." http://www.earthobservations.org/documents/10-Year Implementation Plan.pdf.

Huete, a, K Didan, T Miura, E.P Rodriguez, X Gao, and L.G Ferreira. 2002. "Overview of the Radiometric and Biophysical Performance of the MODIS Vegetation Indices." *Remote Sensing of Environment* 83 (1-2): 195–213..

Imran, Muhammad, Raul Zurita-milla, and Rolf A De By. 2011. "Integrated Environmental Modeling - An SDI-Based Framework for Integrated Assessment of Agricultural Information." In *14th AGILE International Conference on Geographic Information Science*, 1–9. Utrecht, The Netherlands.

Jönsson, Per, and Lars Eklundh. 2004. "TIMESAT—a Program for Analyzing Time-Series of Satellite Sensor Data." *Computers & Geosciences* 30 (8): 833–45.

Kuenzer, Claudia, Marco Ottinger, Martin Wegmann, Huadong Guo, Changlin Wang, Jianzhong Zhang, Stefan Dech, and Martin Wikelski. 2014. "Earth Observation Satellite Sensors for Biodiversity Monitoring: Potentials and Bottlenecks." *International Journal of Remote Sensing* 35 (18): 6599–6647.

Lopez-Pellicer, Francisco J., Walter Rentería-Agualimpia, Rubén Béjar, Pedro R. Muro-Medrano, and F. Javier Zarazaga-Soria. 2012. "Availability of the OGC Geoprocessing Standard: March 2011 Reality Check." *Computers & Geosciences* 47 (November): 13–19.

MacKenzie, C. Matthew, Ken Laskey, Francis McCabe, Peter F Brown, Rebekah Metz, and Booz Allen Hamilton. 2006. "Reference Model for Service Oriented Architecture 1.0."

Mazzetti, P., S. Nativi, and J. Caron. 2009. "RESTful Implementation of Geospatial Services for Earth and Space Science Applications." *International Journal of Digital Earth* 2 (sup1): 40–61.

Meng, Xiaoliang, Fuling Bian, and Yichun Xie. 2009. "Geospatial Services Chaining with Web Processing Service." In *Proceedings of the International Symposium on Intelligent Information Systems and Applications (IISA'09)*, 6:007–010.

Meng, Xiaoliang, Yichun Xie, and Fuling Bian. 2010. "Distributed Geospatial Analysis through Web Processing Service: A Case Study of Earthquake Disaster Assessment." *Journal of Software* 5 (6): 671–79.

Müller, Matthias, Lars Bernard, and Daniel Kadner. 2013. "Moving Code – Sharing Geoprocessing Logic on the Web." *ISPRS Journal of Photogrammetry and Remote Sensing* 83 (September): 193–203.

Nash, Edward, Jans Bobert, Karl-otto Wenkel, Wilfried Mirschel, and Ralf Wieland. 2007. "Geocomputing Made Simple: Service-Chain Based Automated Geoprocessing for Precision Agriculture." In *GeoComputation 2007*. Maynooth, Ireland.

Overdick, Hagen. 2007. "The Resource-Oriented Architecture." *2007 IEEE Congress on Services (Services 2007)*, July. Ieee, 340–47.

Schaeffer, Bastian, Bastian Baranski, Theodor Foerster, and Johannes Brauner. 2012. "A Service-Oriented Framework for Real-Time and Distributed Geoprocessing." In *Geospatial Free and Open Source Software in the 21st Century*, edited by Erwan Bocher and Markus Neteler, 3–20. Berlin, Germany: Springer Berlin Heidelberg.

Schut, Peter. 2007. *OpenGIS Web Processing Service. Open Geospatial Consortium*. Open Geospatial Consortium Inc. http://www.opengeospatial.org/standards/wps.

Shvidenko, Anatoly, Dmitry Schepaschenko, Anatoly Sukhinin, Ian Mccallum, and Shamil Maksyutov. 2011. "Carbon Emissions from Forest Fires in Boreal Eurasia between 1998-2010." In *5th International Wildland Fire Conference*. Sun City, South Africa.

Sun, Alexander. 2013. "Enabling Collaborative Decision-Making in Watershed Management Using Cloud-Computing Services." *Environmental Modelling & Software* 41 (March). Elsevier Ltd: 93–97.

Traore, Abdoul, Philippe Ciais, Nicolas Vuichard, Natasha McBean, Cecile Dardel, Benjamin Poulter, Shilong Piao, et al. 2014. "1982–2010 Trends of Light Use Efficiency and Inherent Water Use Efficiency in African Vegetation: Sensitivity to Climate and Atmospheric CO2 Concentrations." *Remote Sensing* 6 (9): 8923–44.

Tucker, Compton J. 1979. "Red and Photographic Infrared L , Lnear Combinations for Monitoring Vegetation." *Remote Sensing of Environment* 8: 127–50.

Verbesselt, Jan, Rob Hyndman, Glenn Newnham, and Darius Culvenor. 2010. "Detecting Trend and Seasonal Changes in Satellite Image Time Series." *Remote Sensing of Environment* 114 (1). Elsevier B.V.: 106–15.

Verbesselt, Jan, Rob Hyndman, Achim Zeileis, and Darius Culvenor. 2010. "Phenological Change Detection While Accounting for Abrupt and Gradual Trends in Satellite Image Time Series." *Remote Sensing of Environment* 114 (12). Elsevier B.V.: 2970–80..

Vitolo, Claudia, Yehia Elkhatib, Dominik Reusser, Christopher J.a. Macleod, and Wouter Buytaert. 2015. "Web Technologies for Environmental Big Data." *Environmental Modelling & Software* 63 (January). Elsevier Ltd: 185–98.

Warmerdam, Frank. 2008. "The Geospatial Data Abstraction Library." In *Open Source Approaches in Spatial Data Handling*, edited by G.Brent Hall and MichaelG. Leahy, 2:87–104. Advances in Geographic Information Science. Springer Berlin Heidelberg.

Weerawarana, Sanjiva, Francisco Curbera, Frank Leymann, Tony Storey, and Donald F Ferguson. 2005. *Web Services Platform Architecture: SOAP, WSDL, WS-Policy, WS-Addressing, WS-BPEL, WS-Reliable Messaging and More*. Upper Saddle River, NJ, USA: Prentice Hall PTR.

Wosniok, Christoph, Felix Bensmann, Roman Wössner, Jörn Kohlus, and Rainer Roosmann. 2014. "Enriching the Web Processing Service." In *EGU*. Vol. 16. Vienna, Austria.

Wu, Huayi, Lan You, Zhipeng Gui, Shuang Gao, Zhenqiang Li, and Jingmin Yu. 2014. "FAST: A Fully Asynchronous and Status-Tracking Pattern for Geoprocessing Services Orchestration." *Computers & Geosciences* 70 (September). Elsevier: 213–28.

Zhao, Peisheng, Theodor Foerster, and Peng Yue. 2012. "The Geoprocessing Web." *Computers & Geosciences* 47 (October): 3–12.

Free and Open Source Software for Geospatial (FOSS4G) Conference Proceedings

Volume 16 *Bonn, Germany* Article 10

2016

Modelling Population Interactions: the PySAL and Open Source Way

Lorraine Barry
Queen's University Belfast, United Kingdom

Recommended Citation
Barry, Lorraine (2016) "Modelling Population Interactions: the PySAL and Open Source Way," *Free and Open Source Software for Geospatial (FOSS4G) Conference Proceedings*: Vol. 16 , Article 10.
DOI: https://doi.org/10.7275/R5TD9VHM
Available at: https://scholarworks.umass.edu/foss4g/vol16/iss1/10

Modelling Population Interactions: the PySAL and Open Source Way

Lorraine Barry

Queen's University Belfast, United Kingdom – l.barry@qub.ac.uk

KEYWORDS: PySAl; regionalization;geovisualization; spatial interaction modelling; open source geospatial; python

ABSTRACT:

The poster aims to present the use, modification and development of open source geospatial libraries to cluster, visualise and model the flow of movements by commuters in a more meaningful way. The research focus' on the regionalization of census boundaries and aggregated commuter flows to produce consistent commuter regions for spatial interaction modelling. More intuitive boundaries produce models which help reflect influences of demographics (gender, religion, status). The research is based on the three strands of regionalization, geo-visualization and gravity modelling.

Modelling Population Interactions: the PySAL and Open Source Way

Lorraine Barry
PhD Research in GI Science
Queen's University Belfast
l.barry@qub.ac.uk

Abstract

This research outlines the main limitations of the use of current aggregate data for the analysis of commuting flows using the example of Northern Ireland. Current aggregations at ward level vary greatly in size and internal commuting flows and are therefore a highly unsuitable geography for the basis of spatial interaction modelling and the understanding of commuting patterns based on socio-demographic population characteristics.

This research aims to create data driven functional regions to investigate population variables which influence patterns of movement. Such research investigates interactions at regional and local scales; efficiently displays voluminous flows and patterns; and demonstrates applicability to a wider academic, census and industry setting. Creating regionalization geographies of consistent internal flows for modelling, visualising voluminous and complex data flows, and regression modelling of population characteristics can only be attempted by a move to open source developments and adaptation. Python libraries in particular PySAL* provide an invaluable platform to custom create a maximum number of zones with consistent commuter flows. QGIS offers a customisable platform for clear visualization of bulk directional movements. R programming provides a flexible structure to model interactions and demographics.

Why Northern Ireland? Issues of segregation, gender and religion feature heavily within the province. Are these related to observed commuting patterns? The ability to properly model these issues through open source developing methods are important for understanding and evaluation and are demonstrated through the presentation of custom maps, scripts and statistics.

Background to Research

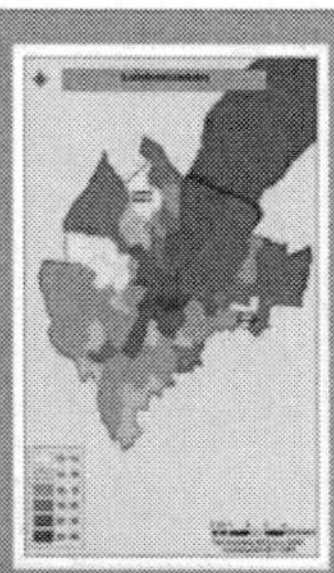

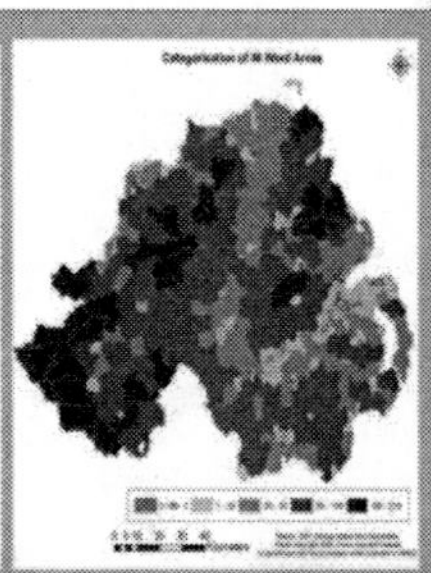

Northern Ireland Wards – Variation in Size and Internal Flows

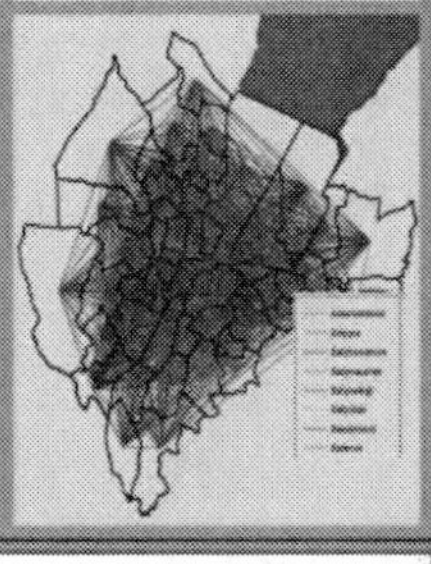

MAUP EF
Small Cell
Scale Unit

Methods

Objectives:
- To seek out more applicable functional areas / regions for interaction data.
- To source and investigate methods for the regionalization of interaction data.
- To draw out and investigate the key population variables which influence patterns of movement.
- To investigate patterns at a regional scale and a local community scale.
- To employ interaction data from Northern Ireland to serve as a case study to model methods, theories and outcomes which are applicable to national levels.

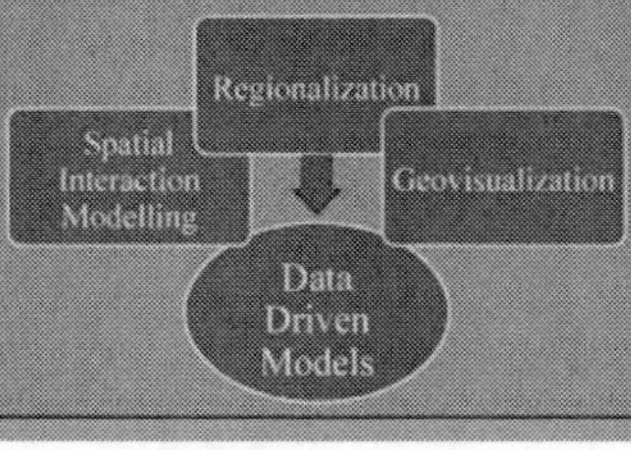

Regionalization

Goal to obtain geography of consistent internal flows for modelling

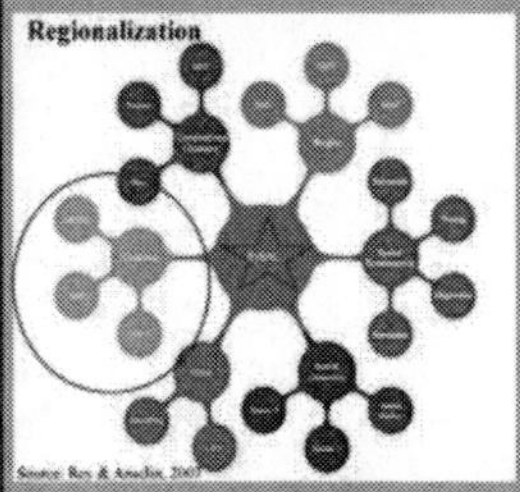

Regionalisation is the clustering or grouping of spatial units into spatially contiguous regions whilst maximising a particular objective function.
(Guo, 2009).

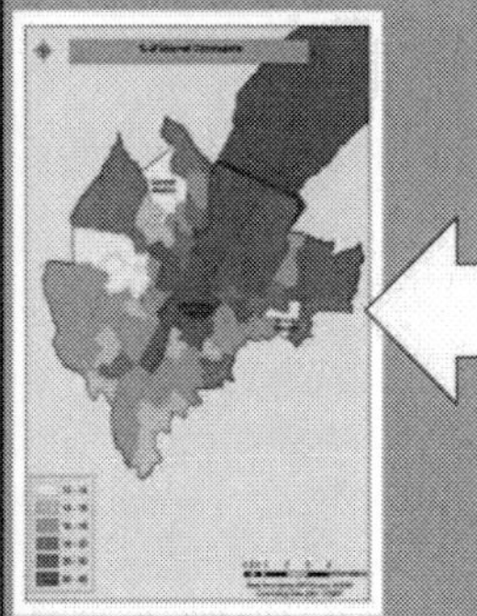

- Creation of purpose-specific zones
- Zones of consistent internal flows
- Generalisations from voluminous datasets
- Diminish effects of irregular zones

PySAL

Region Building Code

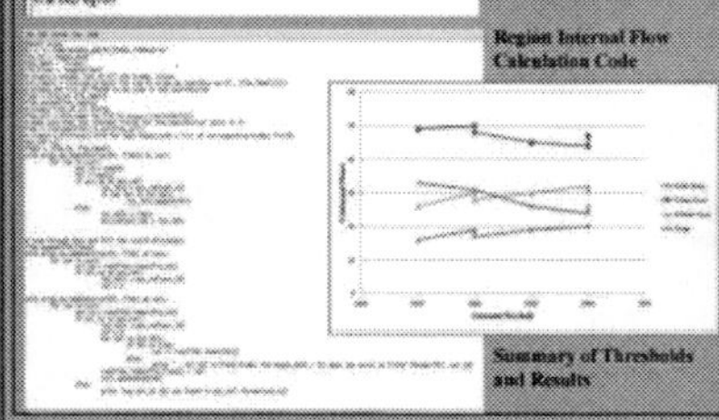

Region Internal Flow Calculation Code

Summary of Thresholds and Results

Spatial Interaction Modelling

SIM - a technique to evaluate the patterns between volume of flows and the underlying socio-economic tendencies of the origin and destination zones.
(Lloyd et al, 2011)

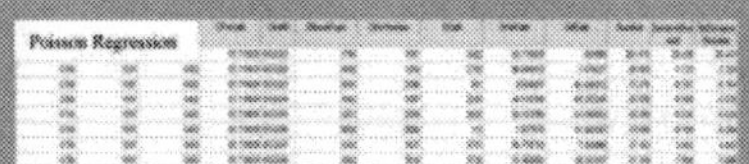

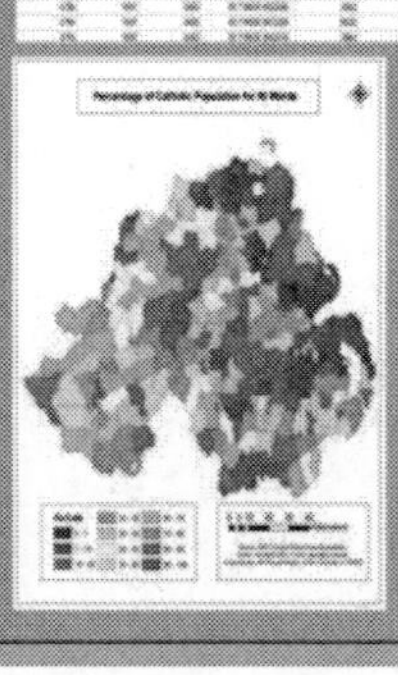

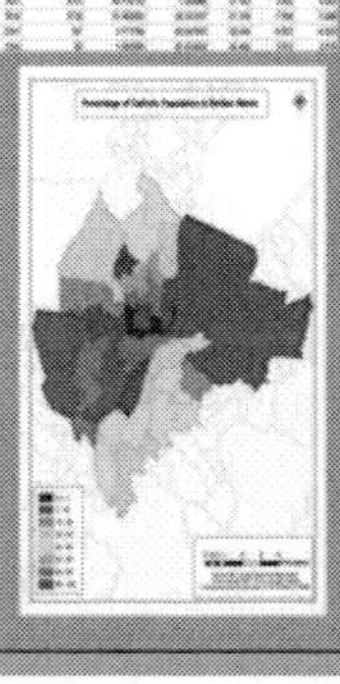

Geovisualization

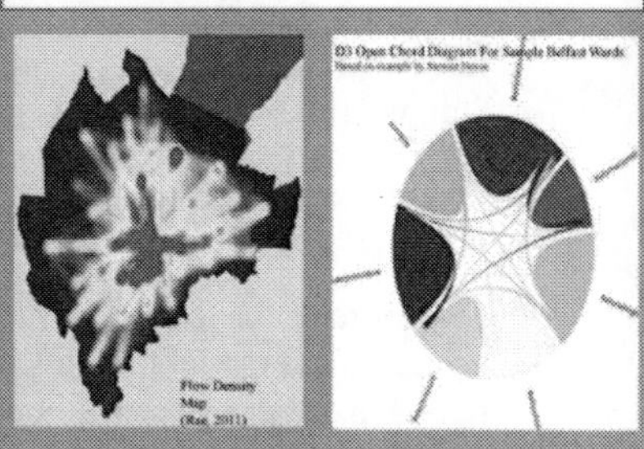

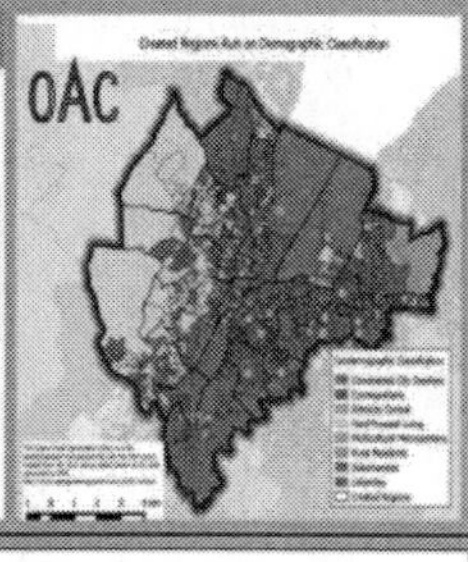

Summary

Choice of Regionalization Method and Objective Function:
Max-P is the algorithm of primary focus for the intelligent aggregation of wards into regions. Max-p is python based and therefore benefits from being open source and highly customisable. It does not depend on a predefined number of regions so therefore the regionalization is driven by the data rather than an enforced arbitrary number of output groupings. (Spielman & Folch, 2015). The aim of this method is to reduce the amount of aggregation necessary to satisfy usability of the data and to create a maximum number of regions.

As part of any regionalization algorithm, the objective function is crucial to output zonings. The choice of Max-p algorithm through the python library PySAL is important as it minimises the amount of aggregation and also allows flexibility in the structuring of the objective function. Currently the default objective function with Max-p is to optimize a homogeneity measure – the total sum of squared deviations SSD. At present results show that created regions have inconsistent proportions of flows largely due to the objective function. The aim of the next crucial stage of research is to modify the associated objective function to optimise for the aims of this research; that is to create flow consistent regions for spatial interaction modelling. Although challenging to adopt an objective function, Spielman & Folch propose that in combination with Max-p algorithm a modification of the objective function is possible to evaluate and shape regions (Spielman & Folch, 2015).

The three strands of this research – **Regionalization, Geovisualization and Spatial Interaction Modelling** – are vital to understanding population flows and only possible through an adoption and development of open source methodologies. The methods presented are largely applicable to any aggregate analysis of interaction data - commuting, migration, and travel and consumer patterns. The use of both python and new or improved regionalization and visualization methods are being actively applied to the study of lifestyle data (Folch & Spielman 2014)

References

Free and Open Source Software for Geospatial (FOSS4G) Conference Proceedings

Volume 16 *Bonn, Germany* Article 11

2016

Shapefile Projection Finder

Manfred Egger
egger-gis.at

Shapefile Projection Finder

Manfred Egger[a, *]

[a] GIS Leisure Project, http://egger-gis.at
manfred@egger-gis.at

KEY WORDS: Road change detection, GPS log, Big data, Digital map building

ABSTRACT:
A new way to find and define the coordinate system of GIS data automatically is presented. This tool is for people without special knowledge of coordinate systems, who want to integrate GIS data with unknown projection to their GIS project.

* Corresponding author

GIS connects our world: From an Austrian to a global solution...

SHAPEFILE PROJECTIONFINDER by Manfred Egger

A new way to find and define the coordinate system of GIS data automatically

Contact: Manfred Egger, Phone: 004369981372857,
Email: manfred@egger-gis.at, Website: www.egger-gis.at

INTRODUCTION AND OVERVIEW:

For The Austrian Service for Torrent and Avalanche Control a tool was developed which can automatically find and define the unknown projection of GIS data for ten Austrian coordinate systems. Since the solution works the support time for problems with gis data and unkown coordinate decreased rapidly.
Here a solution which integrates all projected coordinate systems based on world-wide registered EPSG-Codes will be presented.
GIS-Users with almost no knowledge of coordinate systems and projections have often problems to integrate gis data in their projects, because the datasets have a wrong or no projection definition. Specialized knowledge about coordinate systems and projections is necessary to integrate data fast in a existing gis project.
In Figure 1 is an example where gis data is in the wrong country.

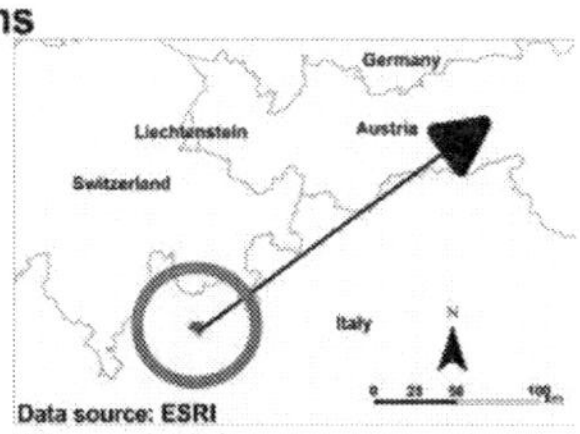

Figure 1: Unknown projection.

PRESUMPTIONS FOR A GLOBAL SOLUTION:

- all places in the world have unique names and latitude/longitude coordinates
- GIS data of the same location can be saved in different coordinate systems

So with the combination of geographic name, latitude/longitude and projected coordinates it is possible to almost automatically detect and determine the unknown coordinate system of the GIS data.
There are different ways to develop such software solutions:
- changing the coordinate system of the gis project until the unknown data is displayed in correct position.
- using names and polygons of adminstrative units (for instance a district)
- distance between latitude/longitude and a sample coordinate of unkown gis data, ...

OPEN SOURCE SOLUTION:

A proposal for a solution was developed with this open source projects basing on Java and NetBeans IDE 8.0.2:
- projfinder.com (open source project by Aaron Racicot)
- geonames.org
- geotools.org (2.7.0)
- spatialrefrence.org

The relevant parameter is the distance value, which describes the distance between latitude/longitude and the center coordinate of the GIS data of unknown origin (look at figure two and three).

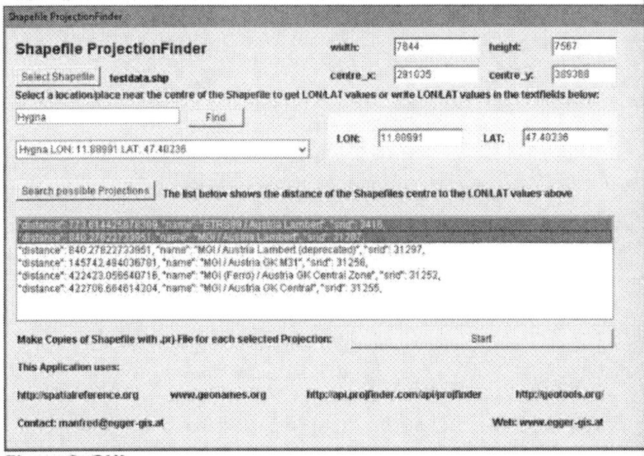

Figure 2: GUI.

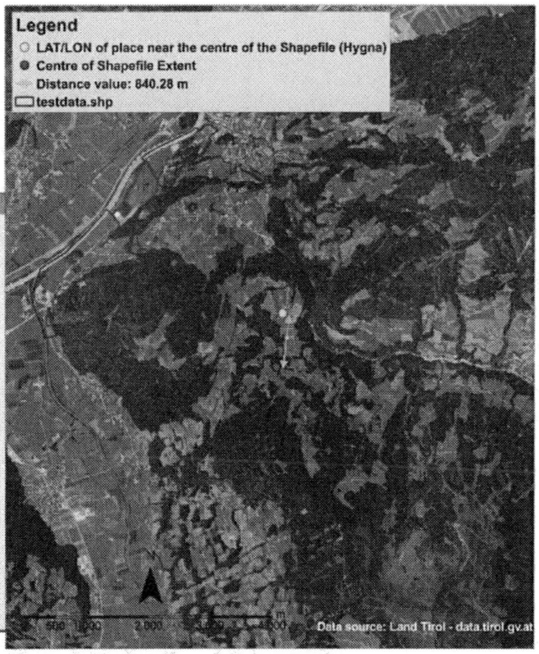

Figure 3: Explanation of distance value.

CONCLUSIONS:

A first proposal to find and define the coordinate system of GIS data automatically was presented. Applications like SHAPEFILE PROJECTIONFINDER help gis users to integrate gis data to their projects fast.
But not all problems can be solved: CUSTOM-Projections, gis data with geographic coordinates or data in print units will make problems. Maybe there are problems with coordinate systems basing on different reference ellipsoids (for instance UTM 32 N (WGS84 and ETRS98)). For this situations it is possible to make copies of gis data and visually select the correct coordinate system in the gis project.

SOURCE CODE:

Java Source Code with licence information: https://github.com/maegger/
After FOSS4G 2016 a free downloadlink of the tool will be published under: http://www.egger-gis.at/

Free and Open Source Software for Geospatial (FOSS4G) Conference Proceedings

Volume 16 *Bonn, Germany* Article 12

2016

Open Source Geospatial Applications to support River Basin Management in Kharaa River Basin, Mongolia

Jürgen Hofmann
Leibniz Institute of Freshwater Ecology and Inland Fisheries

Vanessa Bremerich
Leibniz Institute of Freshwater Biology and Inland Fisheries, Berlin

Jonas Gröning
Leibniz Institute of Freshwater Biology and Inland Fisheries, Berlin

Marc Janssen
terrestris GmbH

Hinrich Paulsen
terrestris GmbH

Recommended Citation

Hofmann, Jürgen; Bremerich, Vanessa; Gröning, Jonas; Janssen, Marc; and Paulsen, Hinrich (2016) "Open Source Geospatial Applications to support River Basin Management in Kharaa River Basin, Mongolia," *Free and Open Source Software for Geospatial (FOSS4G) Conference Proceedings*: Vol. 16 , Article 12.
DOI: https://doi.org/10.7275/R5V9868H
Available at: https://scholarworks.umass.edu/foss4g/vol16/iss1/12

Open Source Geospatial Applications to support River Basin Management in Kharaa River Basin, Mongolia

Jürgen Hofmann[a*], Vanessa Bremerich[a], Jonas Gröning[a], Marc Jansen[b] & Hinrich Paulsen[b]

[a] Leibniz Institute of Freshwater Biology and Inland Fisheries, Berlin, [b] terrestris GmbH – j.hofmann@igb-berlin.de, bremerich@igb-berlin.de, groening@igb-berlin.de, jansen@terrestris.de, paulsen@terrestris.de

KEY WORDS: River Basin Management, Risk assessment, Water Quality Monitoring, Mongolia

ABSTRACT:
An open access geoportal, developed using open source technologies, delivers a comprehensive overview of all geodata available for the Kharaa River Basin in Mongolia. The Water Quality Monitoring Database is made available in PostgreSQL/PostGIS format and embedded in the geoportal. Web maps in the Geoportal can be linked to monitoring data using interactive queries to the Water Quality Database enabling water managers to create thematic maps on demand for the development of a comprehensive river basin management plan for the Kharaa River Basin.

Open Source Geospatial Applications to support River Basin Management in Kharaa River Basin (KRB), Mongolia

Jürgen Hofmann[1], Vanessa Bremerich[2], Jonas Gröning[3], Marc Jansen[4] & Hinrich Paulsen[5]
Contact: Leibniz Institute of Freshwater Biology and Inland Fisheries, Müggelseedamm 310, D-12587 Berlin, +49-30-63924073, email: 1 bremerich@igb-berlin.de, 3 groening@igb-berlin.de, terrestris GmbH, Kölnst. 99, D-53111 Bonn, +49-228-96289951, 4 jansen@terrestris.de, paulsen@terrestris.de

Location of Kharaa River Basin in Mongolia

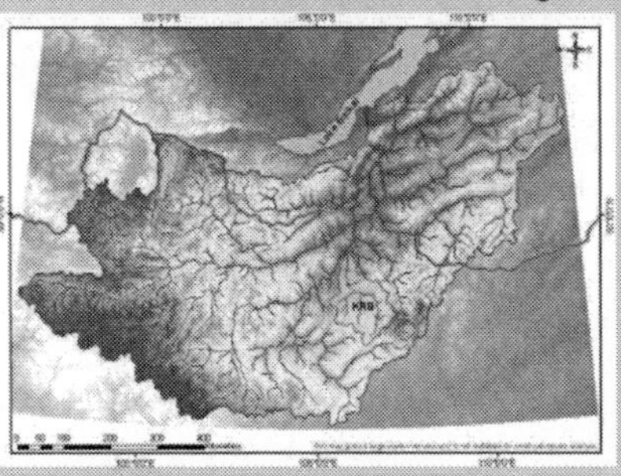

The MoMo Geoportal as open access gateway to support river basin management

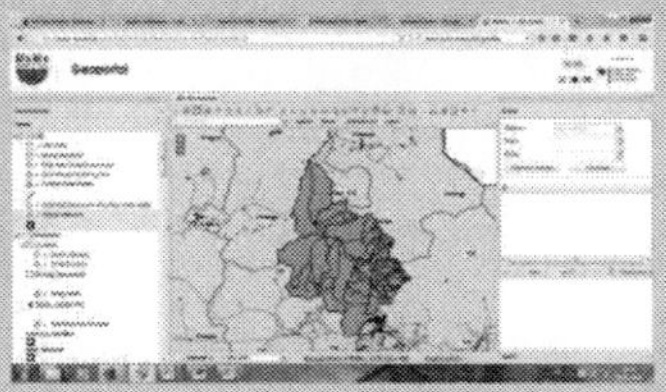

Risk assessment map for ground water bodies linked to the Water Quality Monitoring Database in the MoMo Geoportal

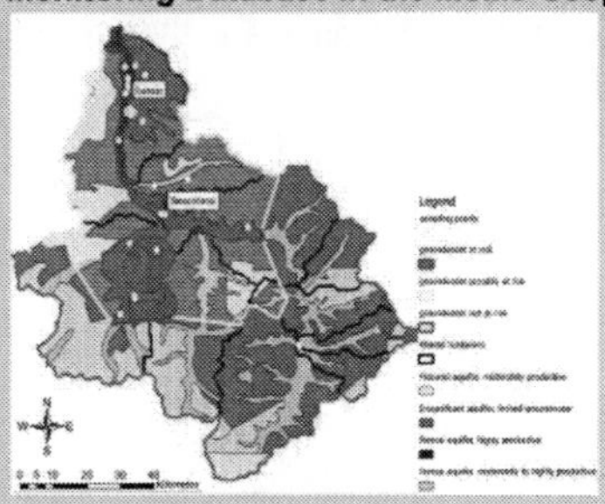

Groundwater bodies of Kharaa River Basin

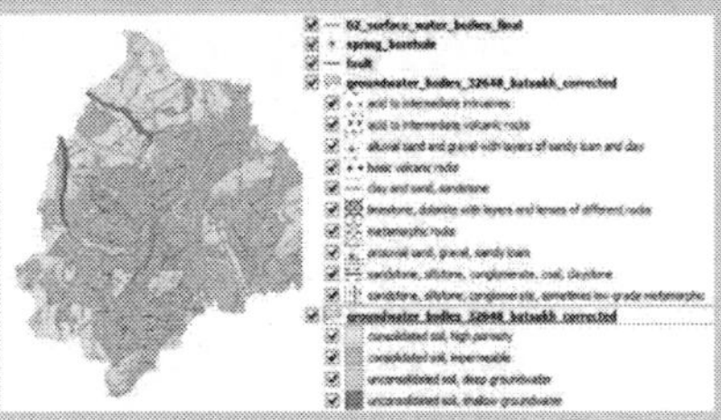

Creating the value chain with Open Source Geospatial Applications

a)The open access MOMO Geoportal (www.iwrm-momo.de/geoportal), developed using open source technologies, delivers a comprehensive overview of all geodata available for Kharaa River Basin

b)The Water Quality Monitoring Database is made available in PostgreSQL/PostGIS format and embedded in the geoportal

c)Maps in the MoMo Geoportal can be linked to monitoring data using interactive queries to the Water Quality Database

d)In that way, Water managers are enabled to create thematic maps on demand for the development of a comprehensive river basin management plan for the Kharaa River Basin

EXAMPLE MAP:

The preliminary pressure / impact analysis indicates ground water bodies in Kharaa River Basin to be

- at risk
- possibly at risk
- not at risk

of failing the environmental objectives for Arsenic in 2013.

The analysis focused on anthropogenic pressures resulting from point and diffuse source pollution and alterations of **groundwater** for the following parameters:

- Nutrient emissions
- Hazardous substances emissions (e.g. heavy metals, CN⁻)
- Pollution indicators (Bor, Chloride)
- Organic pollution

Main contamination sources in Kharaa River Basin:

- Mining activities
- Domestic and industrial wastewater, ash deposits of power plants
- Households without connection to wastewater treatment,
- livestock, agriculture

Possible impacts / failure of good state:

- Elevated to high concentrations of heavy metals (e.g. As, Pb, Zn)
- Elevated to high values of general pollution indicators like chloride, boron and electrical conductivity

Interactions between surface and ground water:

- The porous aquifer supplying Darkhan with drinking water is hydraulically very well connected to the Kharaa River
- Thus, the quality of the Kharaa River may strongly influence groundwater quality in the porous aquifer

Joint German-Mongolian research project "Integrated Water Resources Management in Central Asia: Model region Mongolia, MoMo, Phase III" www.iwrm-momo.de

terrestris

IGB Leibniz-Institute of Freshwater Ecology and Inland Fisheries

Free and Open Source Software for Geospatial (FOSS4G) Conference Proceedings

Volume 16 *Bonn, Germany* Article 13

2016

Classification of land use/cover change in the Kanto region, Japan using FOSS4G and Open Data

Nobusuke Iwasaki
National Agriculture and Food Research Institute, Institute

David Spraguea
National Agriculture and Food Research Institute, Japan

Ayaka Onohara
Japan Society for the Promotion of Science

Classification of land use/cover change in the Kanto region, Japan using FOSS4G and Open Data

Nobusuke Iwasaki[a], David Sprague[a], Ayaka Onohara[b]

[a] National Agriculture and Food Research Institute, [b] Japan Society for the Promotion of Science – niwasaki@affrc.go.jp, sprague@affrc.go.jp, aonoa68@gmail.com

ABSTRACT:

In this study, we classified land use/cover change from the 1880's to 2010's in the Kanto Region, Japan. First, we developed a 100m grid land use/cover database of the 1880's with QGIS application and PostGIS database. The data were published through the GitHub Site as an Open Data (CC BY 4.0).

We imported Japanese national Open Data about land use (1970's and 2010's) to the PostGIS database, and carried out a spatial join with land use/cover data of the 1880's. A model-based clustering method was applied to the land use/cover change data that were summarized based on a 1 km by 1 km grid.

The land use/cover change was classified into 29 categories, and we applied multi-dimensional scaling (MDS) analysis to categorize data. Based on MDS analysis, the categories were divided into 7 groups, and each land use/cover change category had a strong relation with distance and direction from central Tokyo and topographical features.

KEYWORDS: Historical map; land use change; Open Data

Classification of land use/cover change in the Kanto region, Japan using FOSS4G and Open Data

Nobusuke IWASAKI (NIAES, niwasaki@affrc.go.jp), David S. SPRAGUE (NIAES) and Ayaka ONOHARA (JSPS Research Fellow, aonoa68@gmail.com)

NARO

OSGeo.JP

◆Abstract

In this study, we classified land use/cover change from the 1880's to 2010's in the Kanto Region, Japan. First, we developed a 100m grid land use/cover database of the 1880's with QGIS application and PostGIS database. The data were published through the GitHub Site as an Open Data (CC BY 4.0).

We imported Japanese national Open Data about land use (1970's and 2010's) to the PostGIS database, and carried out a spatial join with land use/cover data of the 1880's. A model-based clustering method was applied to the land use/cover change data that were summarized based on a 1 km by 1 km grid.

The land use/cover change was classified into 29 categories, and we applied multi-dimensional scaling (MDS) analysis to categorize data. Based on MDS analysis, the categories were divided into 6 groups, and each land use/cover change category had a strong relation with distance and direction from central Tokyo and topographical features.

◆Introduction

- **Historical land use/cover change are valuable information for biodiversity evaluation, disaster management and many other uses. The Rapid Survey Maps (RSM) that were surveyed in the 1880's (early Meiji Era), are the first modern cartographical map series of Japan and important sources of information on traditional land use in early modern Japan.**
- **We developed a grid based land use database using QGIS and PostGIS, and published the database using GitHub and carried out a model-based clustering method to land use/cover change data derived from Open Data.**

◆Material & Methods

Material

- **Land use data**
  - **1880: The Rapid Survey Maps**
  - **1970's and 2010's : National Land Numerical Information (NLNI) Land use fragmented mesh**

Methods

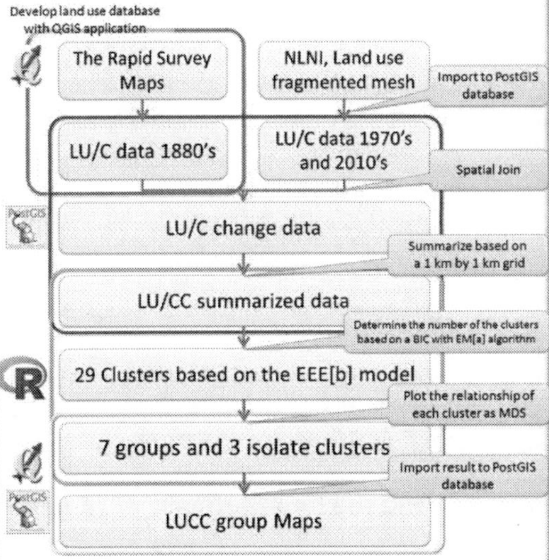

- [a] EM algorithm for maximum likelihood estimation of multivariate mixture models[1].
- [b] EEE models , covariance structure has the following features → distribution: ellipsoidal, volume: equal, shape: equal, orientation: equal

◆Result

- **The relationship between each cluster based on MDS**

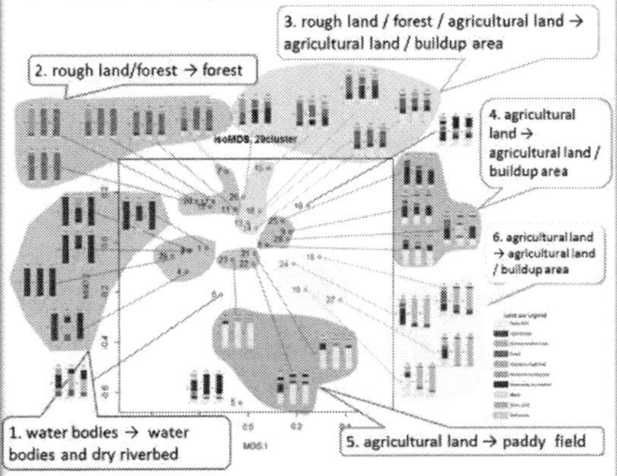

- **Spatial distribution of clusters**

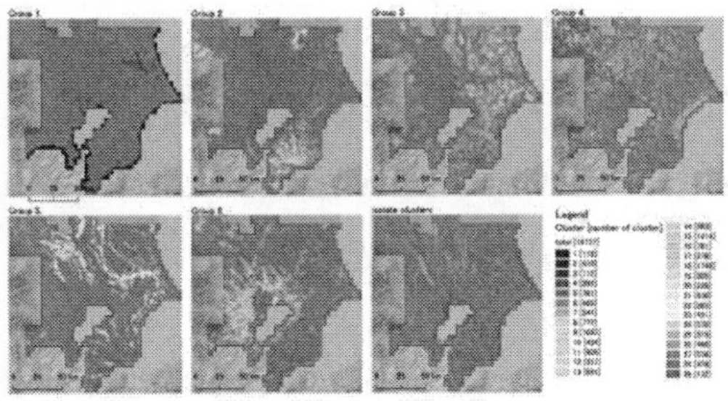

◆Discussion and Conclusion

- **Based on MDS analysis, the categories were divided into 6 groups, and each land use/cover change category had a strong relation to distance from central Tokyo and topographical features.**
  - ➢ **Group 6, i.e. urbanization, occupied any topographical condition nearby Tokyo. On the other hand, Group 5 occupied Alluvial plain far from Tokyo.**
  - ➢ **Land use/cover change to the forest and agricultural land occurred far from Tokyo.**
- **FOSS4G has a good capability to develop and analysis this kind of a large data set.**
  - ➢ **Total numbers of land use/cover change records are about 170 million and land use/cover summarized records are 1.7 thousands.**

- Reference
- [1] Fraley C, Raftery AE. Model-Based Clustering, Discriminant Analysis, and Density Estimation. Journal of the American Statistical Association. 2002 Jun;97(458):611-631

This work was supported by JSPS KAKENHI Grant Number [illegible] and Grant-in-Aid for JSPS Fellows [illegible]

Free and Open Source Software for Geospatial (FOSS4G) Conference Proceedings

Volume 16 *Bonn, Germany* Article 15

2016

Modelling concrete and abstract geoprocessing workflows with ILWIS 4

Rob Lemmens
Faculty of Geo-Information Science and Earth Observation (ITC), University of Twente, Netherlands

Modelling concrete and abstract geoprocessing workflows with ILWIS 4

Rob Lemmens

Faculty of Geo-Information Science and Earth Observation (ITC), University of Twente, Netherlands

– r.l.g.lemmens@utwente.nl

KEYWORDS: Workflows; GIS; Remote Sensing; Geoprocessing; Semantic Web

ABSTRACT:

A workflow represents a combination of process steps to be handled by computers or humans. Geoprocessing workflows consist of geodata (satellite images, in-situ sensor data, human sensor data) and the operations needed for their storage, analysis and presentation. In this research we develop methods for creating and sharing geo-workflows between humans and computers to support knowledge sharing and system interoperability.

Modelling concrete and abstract geoprocessing workflows with ILWIS 4

A workflow represents a combination of process steps to be handled by computers or humans. Geoprocessing workflows consist of geodata (satellite images, in-situ sensor data, human sensor data) and the operations needed for their storage, analysis and presentation. In this research (afrialliance.org) we develop methods for creating and sharing geo-workflows between humans and computers to support knowledge sharing and system interoperability.

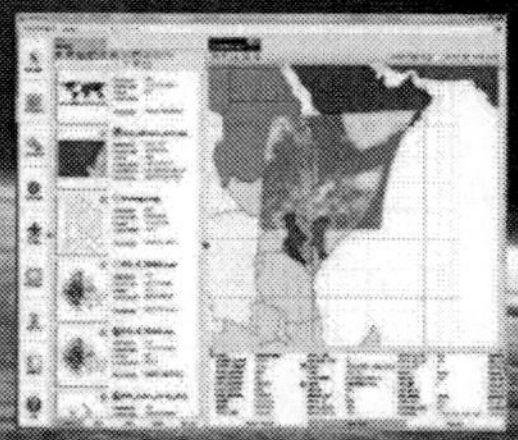

ILWIS 4 Graphical User interface

Concrete workflow

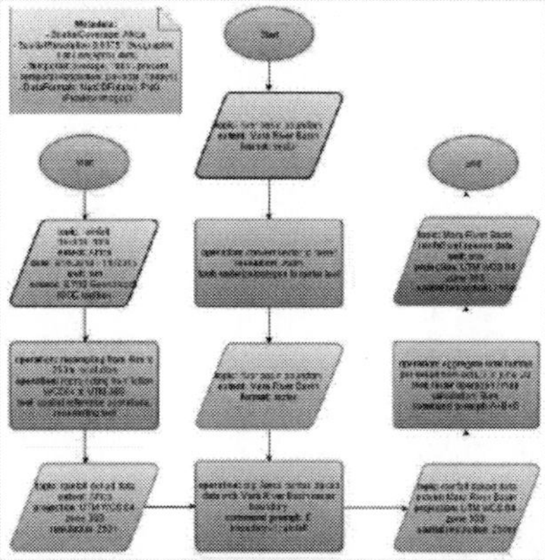

A concrete workflow contains process steps which are executable by a specific software. The visualization represents the system logic. This workflow is executed in ILWIS (52north.org/communities/ilwis) [1] in the MaMaSe project (mamase.org).

Abstract workflow

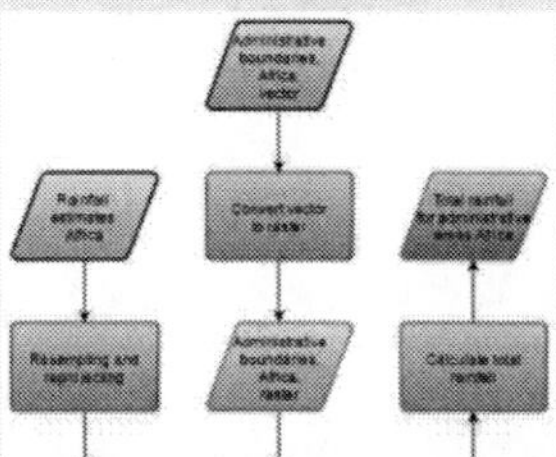

An abstract workflow provides a general overview of process types and their input/output, without necessarily stating data sources and operation parameters. This allows users to create and grasp the general essence of the process steps. An abstract workflow can turn into a concrete workflow and vice versa.

Sharing workflows

Several process languages such as BPMN allow standardized sharing of workflows. We have developed a collaborative web environment using a semantic web-based exchange format in JSON-LD which allows both sharing between machines and humans [2]. Semantically enriched workflows can be visualized in a spatio-temporal explorer based on Linked Data [3].

Right: JSON workflow
Below: ILWIS workflow builder

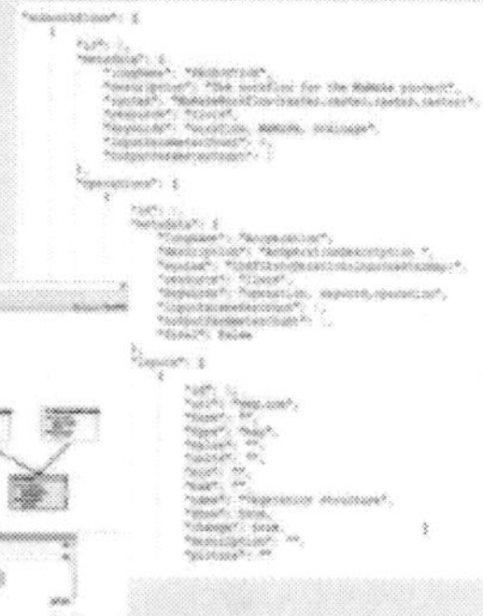

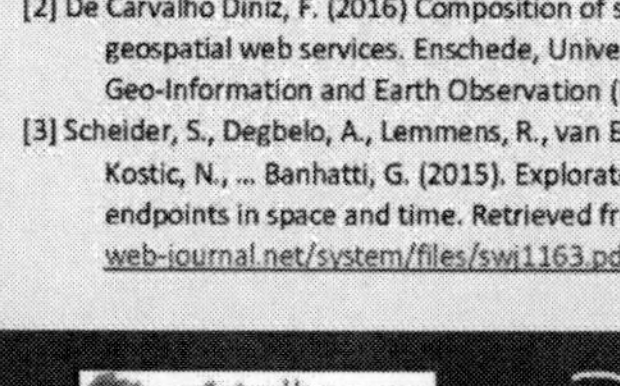

References

[1] ILWIS Open Source GIS/RS software. Video introduction (2016) https://vimeo.com/user29453510/review/153355429/1c1a97df84

[2] De Carvalho Diniz, F. (2016) Composition of semantically enabled geospatial web services. Enschede, University of Twente Faculty of Geo-Information and Earth Observation (ITC), 2016.

[3] Scheider, S., Degbelo, A., Lemmens, R., van Elzakker, C., Zimmerhof, P., Kostic, N., ... Banhatti, G. (2015). Exploratory querying of SPARQL endpoints in space and time. Retrieved from http://www.semantic-web-journal.net/system/files/swj1163.pdf

ITC MaMaSe 52north exploring horizons afrialliance SAXION

Free and Open Source Software for Geospatial (FOSS4G) Conference Proceedings

Volume 16 *Bonn, Germany* Article 16

2016

How to cross a square?

Jakob Miksch
Department of Geography, Heidelberg University, Germany

Stefan Hahmann
Department of Geography, Heidelberg University, Germany

Recommended Citation

Miksch, Jakob and Hahmann, Stefan (2016) "How to cross a square?," *Free and Open Source Software for Geospatial (FOSS4G) Conference Proceedings*: Vol. 16 , Article 16.
DOI: https://doi.org/10.7275/R5F47MBJ
Available at: https://scholarworks.umass.edu/foss4g/vol16/iss1/16

How to cross a square?

Jakob Miksch, Stefan Hahmann

Department of Geography, Heidelberg University, Germany – jakob.miksch@stud.sbg.ac.at, stefan.hahmann@geog.uni-heidelberg.de

ABSTRACT:

The poster compares algorithms how squares can be integrated into a routing engine. A case study showed that most algorithms increase the processing time and the created count of edges significantly.

How to Cross a Square?

Miksch, J. [1,2], Hahmann, S. [1]
[1] Chair of GIScience, Department of Geography, Heidelberg University, Germany
[2] Z_GIS – Department of Geoinformatics, Salzburg University, Austria

1 Introduction

Routing engines have problems to find the shortest path for pedestrians when a square is on the way. The calculated route is often longer than necessary (Map 1). This has a simple reason: Just the boundaries of the squares are integrated into the street network. The interior area of the square is not taken into account.

Map 1: shortest path vs. suggested path by OpenRouteService [1]

2 Method

Creating virtual ways on the square makes it possible to integrate the square into the street network. There are many algorithms for creating these virtual ways. Map 2 until 8 show some examples. The location is the University Square in Heidelberg, Germany. The algorithms have been created with JTS [2]. The blue line shows the new suggested path. The dotted line shows the shortest path.

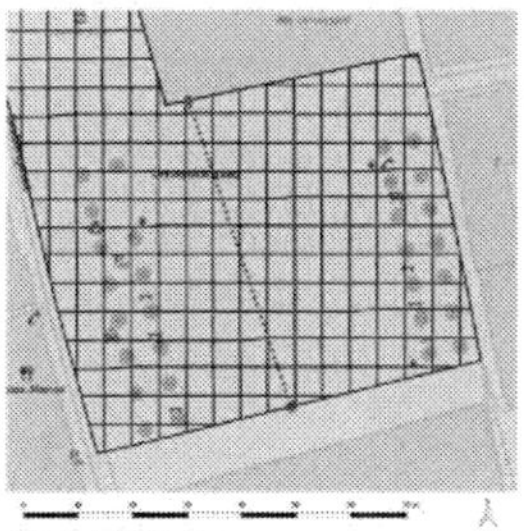

Map 2: Grid 5 meter

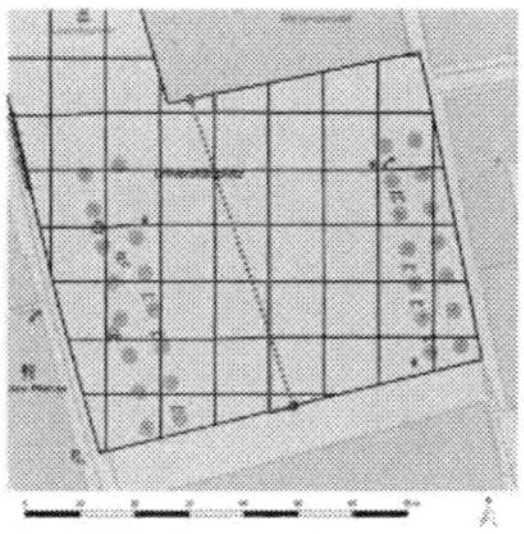

Map 3: Grid 10 meter

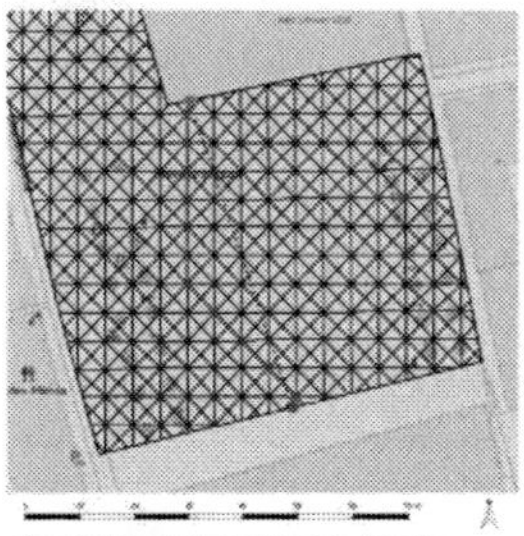

Map 4: Spider grid with diagonals - 5 meter

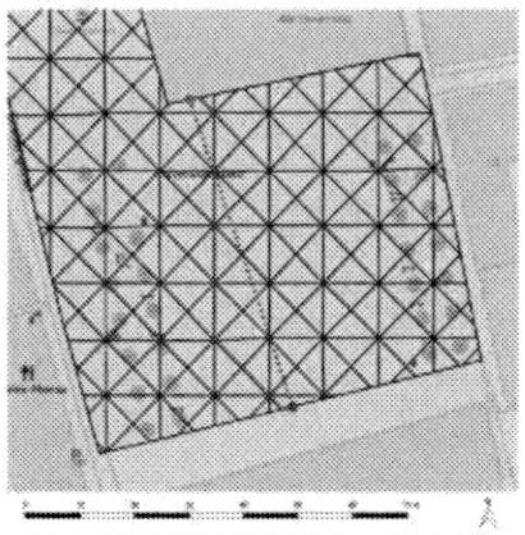

Map 5: Spider grid with diagonals - 10 meter

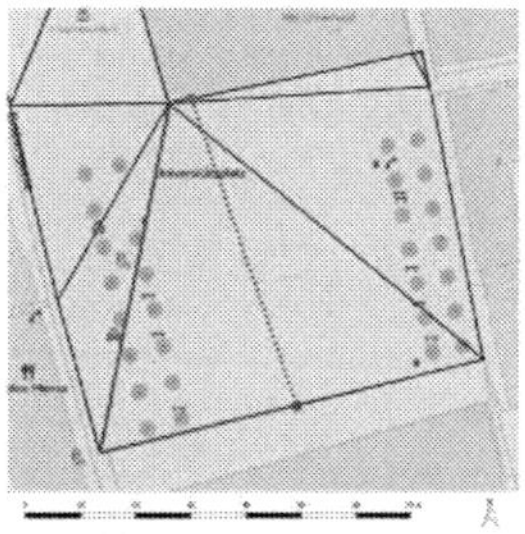

Map 6: Delaunay

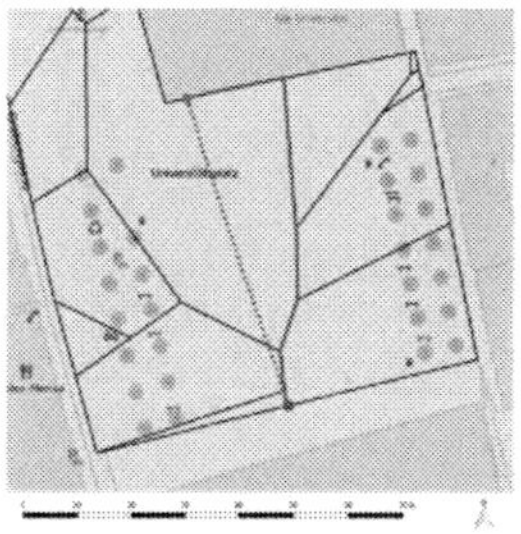

Map 7: Voronoi

3 Comparison of Algorithms

Map 2 and 3 show a grid layer with 5m / 10m grid size. Map 4 and 5 show an extension where two diagonals are added. The algorithms of map 6,7 and 8 take the connection points to the remaining street network into account.

The suggested path of the grids and the "Voronoi" is shorter than in map 1, but has still an unnatural shape. The suggested ways of Delaunay and Visibility are quite long. The reason for it is that there are no field connection points around the red destination node.

In order to compare these algorithms on a larger scale, we calculated them for all accessible squares in Baden-Württemberg using OpenStreetMap [3] data. The processing time and the count of additional edges were compared with a street network created by GraphHopper [4].

The diagram below shows that both parameters become increased significantly by most of the algorithms. So far "Grid 10" seems to be the best compromise. It delivers both an acceptable routing result and does not increase the processing time and the network size to much.

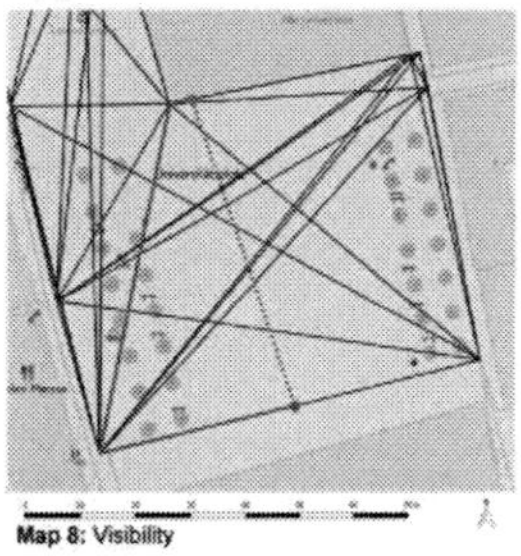

Map 8: Visibility

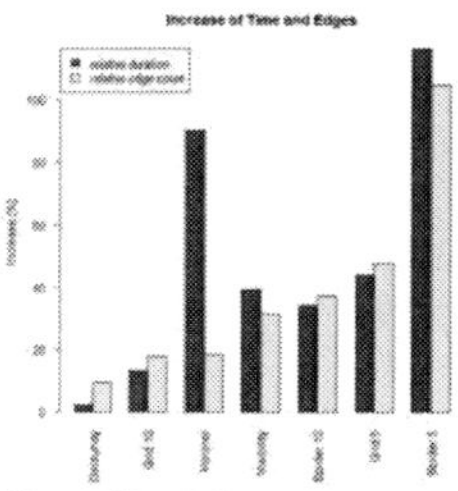

Diagram: Comparison

4 Further Work

This work has done a pre-assessment. In the next steps, the algorithms will be improved in order to make them faster. Moreover some features could be added: The grid could be oriented along the main axis of the square. "Visibility", "Delaunay" and "Voronoi" could be improved by adding more connection points. The final goal is to implement one or many algorithms into a routing engine.

References

[1] www.openrouteservice.org
[2] www.vividsolutions.com/jts/JTSHome.htm
[3] www.openstreetmap.org
[4] www.graphhopper.com
The maps use OpenStreetMap mapnik style as background.

Contact

jakob.miksch@stud.sbg.ac.at
hahmann@uni-heidelberg.de

MyAccessible.EU

A part of the research leading to these results has received funding from the European Community's Seventh Framework Programme (FP7/2007-2013) under grant agreement n° 612096 (CAP4Access)

Free and Open Source Software for Geospatial (FOSS4G) Conference Proceedings

Volume 16 *Bonn, Germany* Article 17

2016

The Construction of Open Data Portal using DKAN for Integrate to Multiple Japanese Local Government Open Data

Toshikazu Seto
Center for Spatial Information Science, the University of Tokyo

Yoshihide Sekimoto
Institute of Industrial Science, the University of Tokyo

Recommended Citation

Seto, Toshikazu and Sekimoto, Yoshihide (2016) "The Construction of Open Data Portal using DKAN for Integrate to Multiple Japanese Local Government Open Data," *Free and Open Source Software for Geospatial (FOSS4G) Conference Proceedings*: Vol. 16 , Article 17.
DOI: https://doi.org/10.7275/R5W957B0
Available at: https://scholarworks.umass.edu/foss4g/vol16/iss1/17

The Construction of Open Data Portal using DKAN for Integrate to Multiple Japanese Local Government Open Data

*Toshikazu Seto [1], Yoshihide Sekimoto [2]

*1: Center for Spatial Information Science, the University of Tokyo, 4-6-1, Komaba, Meguro-ku, Tokyo 153-8505, Japan, 153-8505 Email: tosseto@csis.u-tokyo.ac.jp
2: Institute of Industrial Science, the University of Tokyo

1. Introduction

The classified data categories of Japanese open data are, population and statistical data account for more than 20% of total open data. In other, the geographic data formats (about 30%) are primarily distributed for disaster prevention, education, and tourism sectors, and urban planning diagrams contain positional information regarding facilities. In recent years, the local government and civic tech community in Japan have focused on the context of the FOSS4G. Consequently, the Japanese local government have published open data in more than 200 organizations, but these data are almost simply provided as a file on their website.

In this study, we built open data platform for integrate Japanese local governemnt open data that uses DKAN. DKAN is newer open source portal that extend the CKAN to be used as an Drupal, which feature has been enhanced, such as design template and map viewing.

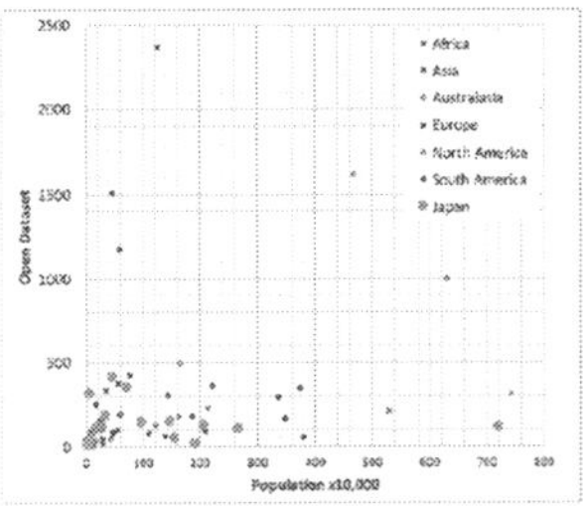

Fukuoka Open Data Portal
http://ckan.open-governmentdata.org/

Fig.1 The Launch Open Data Cities by Using CKAN Platform

Table 1 The Comparison of CKAN and DKAN

Feature	DKAN	CKAN
Language	PHP	Python
Platform	Drupal	Pylons
Database	MySQL, PostgreSQL, SQL Server, or Oracle	PostgreSQL
Geodata Function	Geo File Entity Geofield Map Visualization Entity Geojson Bundle	ckanext-spatial ckanext-geoview ckanext-basemap

2. Collection of Open Data in Japan

Most Japanese open data, it is impossible to harvesting automatically. Because, we need to developed a supporting tool for collecting Japanese open data. The tool is built on Electron (http://electron.atom.io/) framework, it is possible to list the static open data website data with CSV. Electron is build cross platform desktop apps with JavaScript, HTML, and CSS. We also released a full list of meta data (10 categories) in spreadsheet format. Thus, it become possible to consider the status of Japanese local government open data.

Fig. 2 The Opendata-scraping Tool by Electron

Fig. 3 Open Data Metadata List

3. The Construction Open Data Portal using DKAN

DKAN are those that have been implanted in the distribution of Drupal with the architecture of the CKAN. Screen configuration and function customization can be performed on the GUI difficult in CKAN an advantage, it can be complex rights management. The geodata preview function has been enhanced in the initial state, of course, mapping, graphing, can correspond to the open data such as GeoJSON and KML. Therefore, it is possible to use while checking on the Web map before use a variety of geospatial open data.The API function is also supported in RDF/JSON formats, it is possible to display the query sample on the portal site.

We have published to about 7,500 Japanese local government open data. The characteristics of the Japanese of open data is considered that there are many facilities information about the disaster, tourism and education.

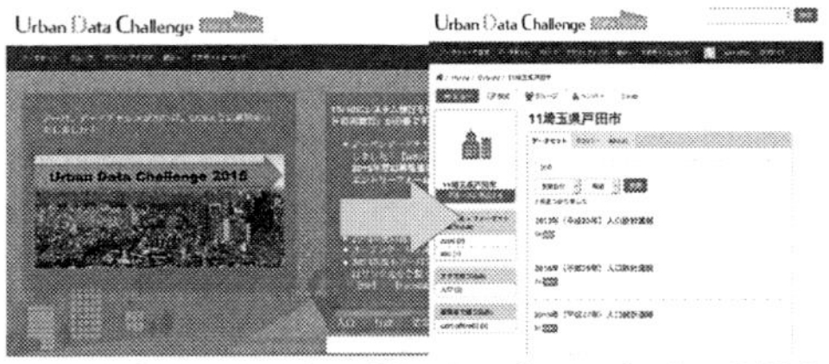

Fig.4 The Construction of User Interface Design by DKAN site
http://udct-data.aigid.jp/

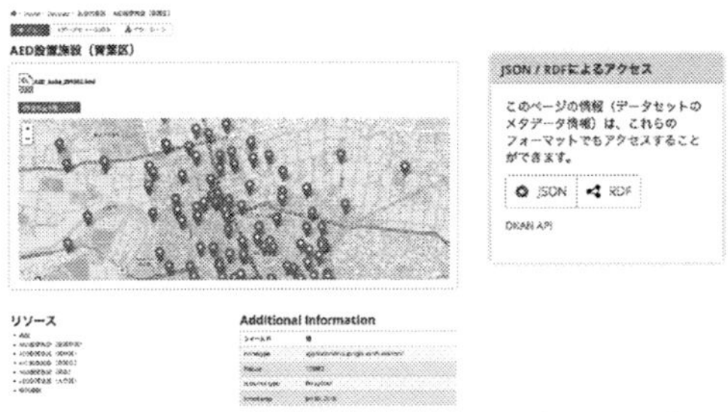

Fig.5 The Map Preview of Geospatial Data (cf. KML) and RDF-API

Table 2 The Top 10 Categories of Open Datasets Type in DKAN site

Data Type	Datasets	Data Type	Datasets
Population	537	Tourism	168
Administration	328	Agriculture	125
Disaster	295	Social Welfare	118
Education	261	Town Planning	108
Infrastructure	169	Transportation	106

4. Conclusions

We have examined the transition of viewing the number from October 1, 2015 to March 31, 2016. The 70% were new visitors, but it was a returning user visited for about 30% to obtain a wide variety of data. DKAN is highly search properties compared to CKAN, API also considered that it was often repeated use of the fact that are substantial.

Fig.5 The Transition of Page View of DKAN site

Free and Open Source Software for Geospatial (FOSS4G) Conference Proceedings

Volume 16 *Bonn, Germany* Article 18

2016

Development of Environment Spatial Information Service Utilizing Mobile Location-Based Technology

Jeongho Yoon
Korea Environment Institute

Recommended Citation

Yoon, Jeongho (2016) "Development of Environment Spatial Information Service Utilizing Mobile Location-Based Technology," *Free and Open Source Software for Geospatial (FOSS4G) Conference Proceedings*: Vol. 16 , Article 18.
DOI: https://doi.org/10.7275/R5X63K4Z
Available at: https://scholarworks.umass.edu/foss4g/vol16/iss1/18

Development of Environment Spatial Information Service Utilizing Mobile Location-Based Technology

Jeongho Yoon

Korea Environment Institute – jhyoon@kei.re.kr

ABSTRACT:

Rapid implementation of mobile devices and significant development of communication technology lead diverse changes over the society at large. Especially, Location-Based Technology of mobile devices is recognized as an important tool to improve task efficiency of tasks that requires on-site investigation and management. The research develops the way to improve environmental administrative work by utilizing Location-Based Technology of mobile devices and environmental geographic information towards tasks for both instruction and inspection of environmental pollutant discharge industries. In addition, it defines function and improvement direction of environment monitoring tasks via test bed implementation to developed mobile application and spatial information database.

KEYWORDS: Environment Monitoring; Mobile; Location-Based; Environmental Geographic Information

Korea Environment Institute

Environment spatial information service utilizing mobile location base technology

1. Research summary

❖ Research Background & Necessity of Development

- Improvement of Location Based Service market
- Need of creditability of public information

Due to Limitation of current environmental information service

Currently, there is lack of task performers' utilization and environmental information service for making structural decision, and also lack of task process to optimize task efficiency

Due to drawbacks of one way delivery of information

Previous service has only provided web system information without mobile optimization

There was lack of providing information of issues like environment influence assessment to public, which is closely related to public safety and national property

Improvement of task system for location based environmental information service

Aims on enhancement of task efficiency as the service rationalize the way and standard of environmental task via location based environmental information service To construct foundation, which connects public information, enables information inquiry and information input

Need of national participation service

There is necessity of building process reflecting consumers requirement based on competent and substantially interesting policy

Which service makes public easy to access to policy by expending and sharing information utilizing mobile devices and social media

❖Research Target

- Service model development of environmental spatial information
- Platform development of environmental spatial information

  - To construct service platform and optimize service model, which suits for previous Environmental Geographic Information System(EGIS, EIASS, etc.) development

- Constructing service platform and model development
- Service App. development of environmental Geographic Information

  - Environment Influence Assessment/Environment monitoring Information service App. development(Task supporting App.)
  - Two way communication App. for environment information(Nationwide service App.)

2. GeoServer based GIS system

❖ GeoServer

- Opensource based GIS software server developed with Java enabling editing and sharing geospatial data

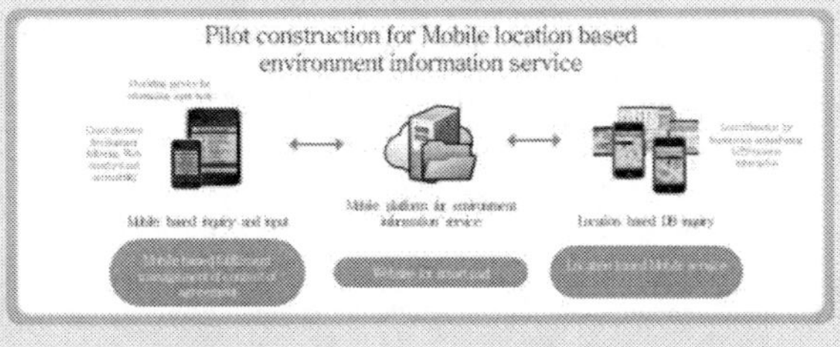

- GeoServer aims on role as a node inside free and open type Spatial Data Infrastructure
- GeoServer performs a role as an Apache HTTP server, which provide open and free type web server for HTML expression, toward spatial data

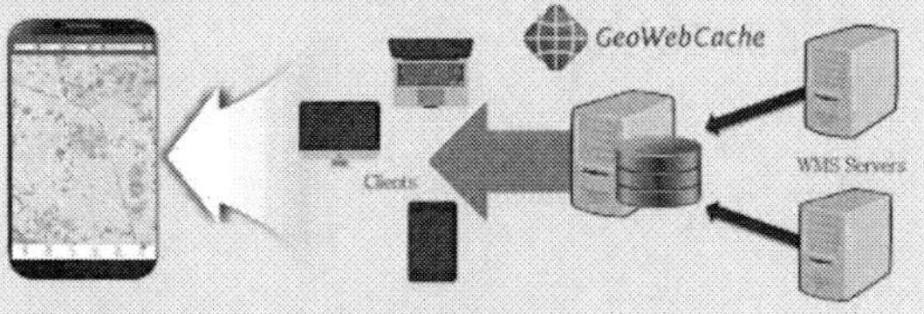

❖ Key features of GeoServer

- Providing user-friendly UI
- No regard to OS (JAVA, WEB Interface based)
- Cache support (GWC, Cache-Control)
- Real-time transformation to diverse coordinates
- Data Processing

3. Research achievement

- Providing data selection and service, and public data analysis utilizing Open API
- Development environment construction accessible via Smart environment(iOS, AndroidOS) by utilizing Open API in relation to environment information
- Development of two way communication App. for environment spatial information

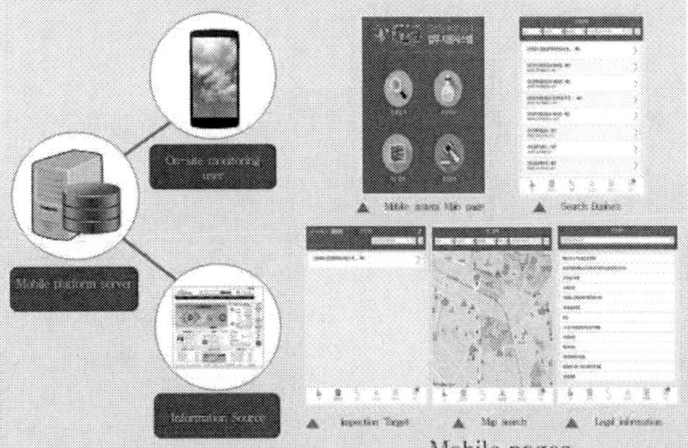

Mobile pages

4. Research application plan

- Replacement of actual image via mobile input of location information
  - Grasp of fulfillment history and date/place information via mobile input of content of agreement
  - Enhancement of record of follow-up management via on-site inspection

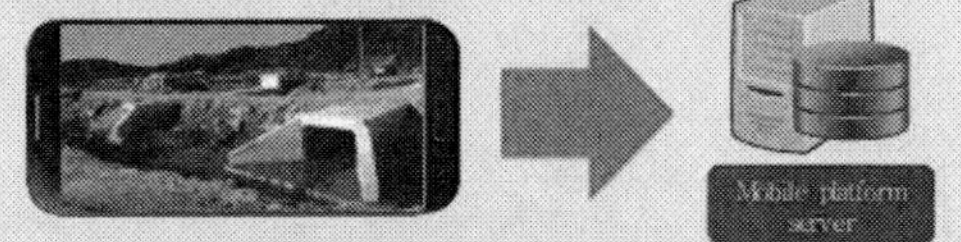

- On-site application of location analysis and inquiry
  - On-site application of location analysis and inquiry App. for environment assessment by utilizing contents of environment assessment information supporting system

Free and Open Source Software for Geospatial (FOSS4G) Conference Proceedings

Volume 16 *Bonn, Germany* Article 19

2016

GRASS in the Desert? Developing FOSS tools for monitoring desertification

Anna Zanchetta
University of Bologna, Italy

Recommended Citation
Zanchetta, Anna (2016) "GRASS in the Desert? Developing FOSS tools for monitoring desertification," *Free and Open Source Software for Geospatial (FOSS4G) Conference Proceedings*: Vol. 16 , Article 19.
DOI: https://doi.org/10.7275/R51V5C55
Available at: https://scholarworks.umass.edu/foss4g/vol16/iss1/19

GRASS in the Desert? Developing FOSS tools for monitoring desertification

Anna Zanchetta

University of Bologna, Italy – anna.zanchetta@unibo.it

ABSTRACT:

The use of Change Vector Analysis (CVA) combined with the Tasselled Cap tranform (TCT) is a powerful remote sensing tool to monitor forests and vegetated areas, but its application to arid and semiarid environment is not straightforward.

This question is tackled through the calculation of a new set of TCT coefficients using R and GRASS-GIS for SPOT and Landsat satellites, then applied and tested in change detection analysis on a short (seasonal) and a long (decades) temporal scale.

Results show that the combined procedure is an effective method to detect changes in desert environment. Furthermore, the new TCT allows the use of this combined procedure for studies in arid and semi-arid regions, eliminating the doubts on its compatibility with the area of study.

Further development is the creation of a new GRASS-GIS module to performe CVA, thus enabling the simple usage of this technique, until now not available in most common software.

KEYWORDS: Desertification; Change detection; Grass-GIS; R

GRASS IN THE DESERT? Developing FOSS tools for monitoring desertification

A. Zanchetta[a], G. Bitelli[a], A. Karnieli[b]

[a] DICAM Dep. of Civil, Chemical, Environmental and Materials Engineering
Bologna University, VialeRisorgimento 2, 40136 Bologna, Italy

[b] The Jacob Blaustein Institutes for Desert Research
Ben Gurion University of the Negev, Sede Boqer Campus, 8499000 Israel

anna.zanchetta@unibo.it

FOSS4G BONN 2016 24th - 26th August

OBJECTIVE

The use of **Change Vector Analysis (CVA)** combined with the **Tasselled Cap Tranform (TCT)** is a powerful remote sensing tool to monitor forests and vegetated areas, but its application to arid and semiarid environment is not straightforward.

This question is tackled through the calculation of a new set of TCT coefficients using **R** and **GRASS-GIS** for SPOT and Landsat satellites, then applied and tested in change detection analysis on a short (seasonal) and a long (decades) temporal scale.

Further development is the creation of a new GRASS module (**i.cva**) to performe CVA, thus enabling the simple usage of this technique, until now not available in most common software.

CASE STUDY

Azraq Oasis (Jordan) is a good case study since it shows a clear seasonal pattern and was subject in the 90s to the complete drying out of the natural springs.

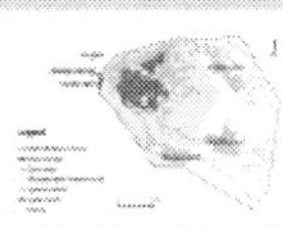

Digitized 1979 map of Azraq Oasis

METHODS

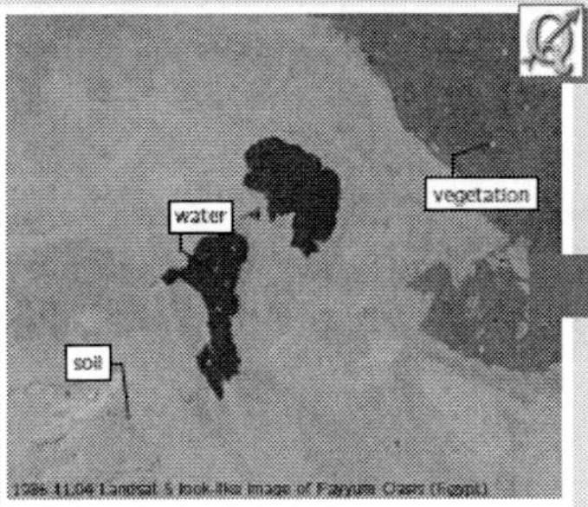

Selection of 20 squares of 100 pixels for three categories of surface cover, for different locations and different seasons of the year (at least 6 images, here shown as an example a location in the Sahara desert in Egypt).

TRANSFORM CALCULATION

Importing the selection in R through Grass, to obtain 2000 pixels for each category.

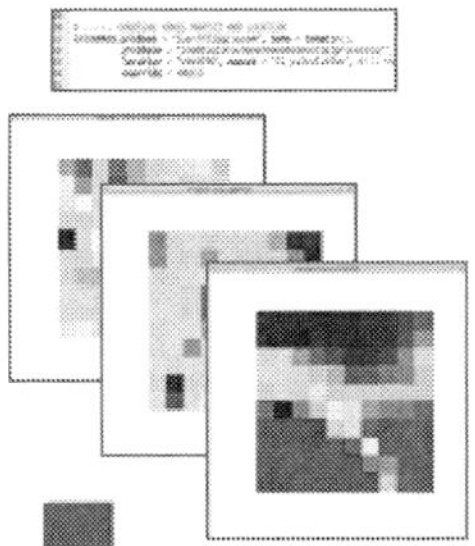

Calculating TCT through adapted Gram Schmidt orthogonalization for each image then obtaining the final transform by average on location and time.

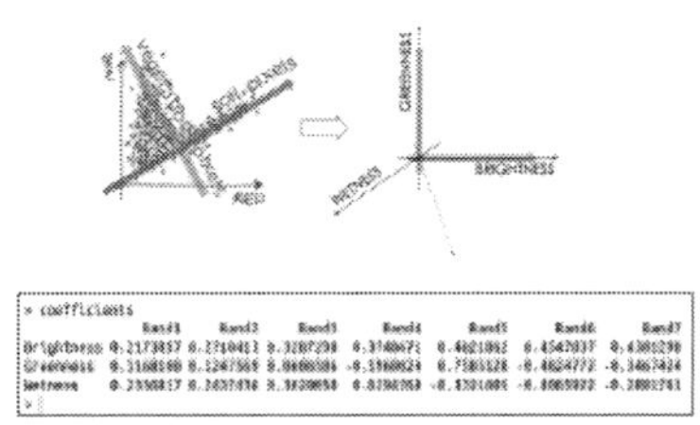

Creating the adapted Brightness and Greenness maps for the CVA analysis (GRASS scripts in **BASH**).

Calculating CVA: ANGLE and MAGNITUDE maps (GRASS scripts).

CHANGE VECTOR ANALYSIS

Obtaining the change detection map by threshold.

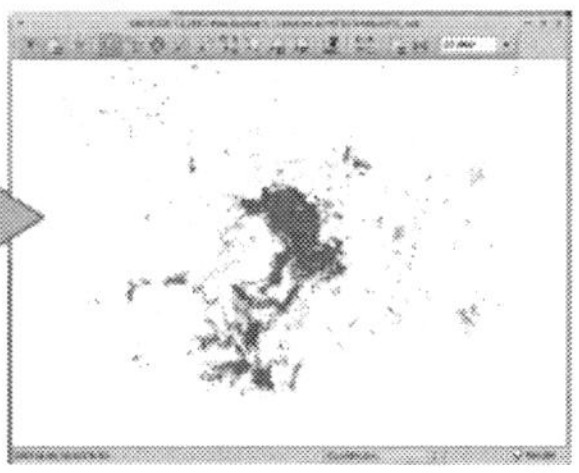

Converting the scripts to a new GRASS module.

RESULTS

SPOT satellites

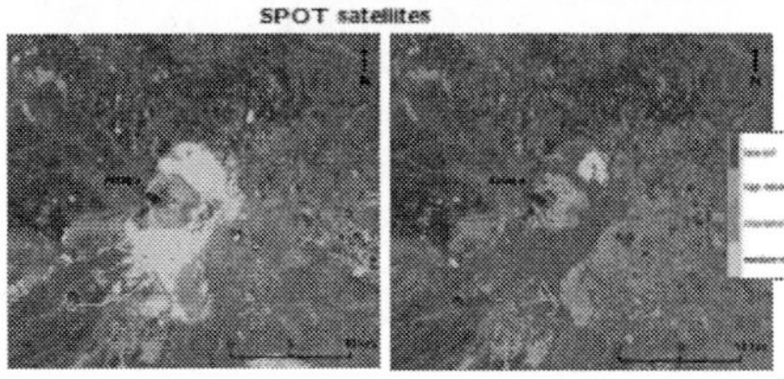

Seasonal change in Azraq Oasis (February – June 2013) detected with the existing TCT [left] and with the new desert-adapted TCT [right].

Landsat satellites

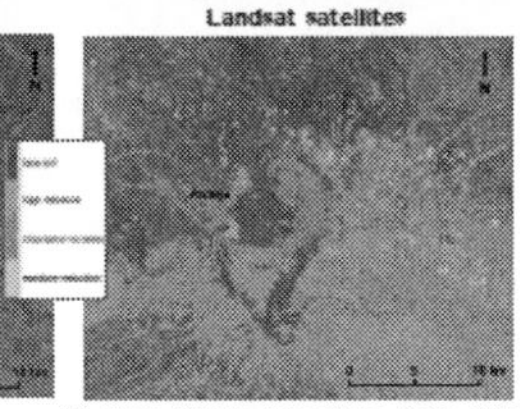

30 years change over Azraq Oasis (1984-2013) during the summer season (August) detected with the new desert-adapted TCT.

CONCLUSION

The Change Vector Analysis (CVA) applied to the new Tasselled Cap Transform (TCT) has been proven to be an effective remote sensing method to detect changes in desert environment.

In the same time, the newly calculated TCT coefficients allow the use of this combined procedure for studies in arid and semi-arid regions eliminating the doubts on their compatibility with the area of study.

The integration of the scripts used for the research into the GRASS environment makes this technique easily accessible for other researchers.

AUTHOR INDEX

Barry, Lorraine 56
Berg, Matthias 1
Bremerich, Vanessa 62
Eberle, Jonas 39
Egger, Manfred 59
Evangelidis, Konstantinos 31
Gröning, Jonas 62
Hahmann, Stefan 71
Hofmann, Jürgen 62
Iwasaki, Nobusuke 65
Janssen, Marc 62
Krause, Detmar 1
Lemmens, Rob 68
Lennert, Moritz 11
Miksch, Jakob 71
Onohara, Ayaka 65
Papadopoulos, Theofilos 31
Paulsen, Hinrich 62
Schwarz, Egbert 1
Sekimoto, Yoshihide 74
Seto, Toshikazu 74
Spraguea, David 65
Vito, Domenico 23
Voinov, Sergey 1
Yoon, Jeongho 76
Zanchetta, Anna 79